SOURCEBOOK OF HYDROLOGIC AND ECOLOGICAL FEATURES

WATER RESOURCE REGIONS OF THE CONTERMINOUS UNITED STATES

Robert M. Cushman
Stephen B. Gough
Mary S. Moran
Robert B. Craig

Environmental Sciences Division
Oak Ridge National Laboratory
Oak Ridge, Tennessee
operated by
Union Carbide Corporation
for the
U.S. Department of Energy

1980

ANN ARBOR SCIENCE
PUBLISHERS INC
P.O. BOX 1425 • ANN ARBOR, MICH. 48106

79 014969

Numerous limnology and hydrology texts have appeared recently. Several present theoretical and factual data on these subjects in a readable and scientifically credible manner, but none provides a systematic description of the water resources of the United States. Useful volumes that cover specific aspects of the nation's aquatic environs (e.g., D. G. Frey 1966)[1] or with particular regional emphasis are available, but the reader interested in obtaining a broader view of the nation's water resources — hydrology, ecology, water use, water quality, and legal environmental status — has had to access a myriad of books, federal and state reports, and journal publications. The reader interested in obtaining information specific to a particular region, on the other hand, often found the process inordinately time-consuming.

This book was written in order to fill a void in the literature relating to the many aspects of our nation's water resources. We hope it will find use in two major areas: (1) in providing an overview of the country's water resources and (2) in serving as a entrée into more detailed investigations of specific related topics. Reference to other publications throughout should provide guidance for more detailed study.

Aside from providing information to academicians, this volume will aid regional planners and others concerned with water resource problems. For example, the National Environmental Policy Act (NEPA) of 1969 requires federal agencies and many federally regulated private organizations to evaluate the consequences of any programs, policies, or legislation that they propose at the earliest practical point in the decision-making process. Many such actions impact broad regions of the United States, for example, those affecting resource utilization or pollution standards. Likewise, many states and coalitions of states participate in regional planning that includes water resource assessment. The descriptions presented here should facilitate such considerations in the future.

Our thanks go to the following technical reviewers: Drs. S. G. Hildebrand, D. D. Huff, and W. Van Winkle of the Environmental Sciences Division, Oak Ridge National Laboratory (ORNL); Ms. E. K. Triegel, Dr. J. E. Dobson, and Mr. A. D. Shepherd of the Energy Division, ORNL; Dr. R. W. Brocksen of the Electric Power Research Institute; and Dr. J. G. Wiener

---

[1]D. G. Frey (Ed.). 1966. *Limnology in North America* (Madison, WI: The University of Wisconsin Press).

of the U.S. Fish and Wildlife Service. We are much indebted
to Ms. L. S. Corrill, Information Division, ORNL, for editorial
assistance in preparation of the manuscript and to Drs. S. I.
Auerbach and D. E. Reichle for support and encouragement.
Funding for preparation of portions of this volume and sug-
gestions as to its content were obtained from Dr. Anthony J.
Dvorak of Argonne National Laboratory and Mr. Steven Frank,
Economic Regulatory Administration, the U.S. Department of
Energy.

This is publication number 1435 of the Environmental
Sciences Division, ORNL, under contract number W-7405-eng-26,
between Union Carbide Nuclear Division and the Department of
Energy.

Robert M. Cushman
Stephen B. Gough
Mary S. Moran
Robert B. Craig

CUSHMAN   GOUGH   MORAN   CRAIG

ROBERT M. CUSHMAN is an aquatic ecologist at Oak Ridge National
Laboratory. He received his MS in Ecology from the University
of Tennessee and his BA in Zoology from Swarthmore College.
Currently in the Environmental Impacts Program within ORNL's
Environmental Sciences Division, he has participaged in assessments of geothermal energy, coal extraction and utilization,
and conversion of coal to synthetic fuels. His scientific
papers are in the fields of aquatic invertebrate ecology and
trace element effects on aquatic biota. He is a member of the
American Association for the Advancement of Science, the American Society of Limnology and Oceanography, the Ecological
Society of America, the North American Benthological Society,
the Societas Internationalis Limnologiae, and the American
Fisheries Society.

STEPHEN B. GOUGH is a research aquatic ecologist with the Environmental Sciences Division at Oak Ridge National Laboratory.
He holds a PhD in Botany, Oceanography and Limnology from the
Unviersity of Wisconsin-Madison, and a BS, magna cum laude,
in Biology and Chemistry from Carroll College. At ORNL he has
participated in the analysis of environmental impacts of portions of the nuclear fission fuel cycle, aspects of the coal
and petroleum fuel cycles, fusion energy, energy storage, and
aquatic biomass systems. His research interests include algal
synecology, autecology and systematics; autotrophic biotech-
nology; the ecology of free-living facultative human pathogens;
and lake restoration/enhancement. He is a Certified Professional Ecologist and has written numerous scientific papers.
He is a member of the American Association for the Advancement
of Science, the American Institute of Biological Sciences, the
American Society of Limnology and Oceanography, the Ecological
Society of America, the Phycological Society of America, the
Society of Sigma Xi, the Water Pollution Control Federation,
and the International Association of Crenobiologists.

MARY S. MORAN is a hydrogeologist with the Environmental Sciences Division at Oak Ridge National Laboratory. Currently a
doctoral candidate in Agricultural Engineering at the University of Tennessee, she earned a MS in Geology and Engineering

from Vanderbilt University and a BS in Geology and Geography
from Tennessee Technological University. She has worked as a
groundwater hydrologist with the U.S. Geological Survey, Water
Resources Division, Tennessee District. Her initial responsi-
bility with the ORNL Energy Division was to assess the hydrogeologic environment of geothermal resource areas and the
hydrogeologic impacts resulting from exploration and development activities. She also developed assessments of the hydrologic impacts resulting from coal mining and processing. With
the Environmental Sciences Division, she is Hydrogeologic
Investigations Coordinator for low-level radioactive waste
disposal. She is the author or coauthor of several publications
in the field of hydrogeology. She is a member of the National
Water Well Association and the Geological Society of America.

ROBERT B. CRAIG is Manager of the Environmental Impacts
Program at Oak Ridge National Laboratory. He received his
PhD in Ecology, and MS and BS degrees in Zoology from the
University of California at Davis.

The Environmental Impacts Program evaluates the environmental consequences of proposed Department of Energy projects,
programs and policies, and reports these for incorporation into DOE planning and decision-making. It deals with all major
energy resource and conversion systems, including nuclear;
coal gasification, liquefaction and direct combustion; oil
and gas; geothermal; solar; and fusion. The program also
provides input for major policy decisions and legislation.

Dr. Craig has published numerous scientific papers and has
coauthored a book on coastal ecology. A Certified Professional Ecologist, he is a member of the Ecological Society of
America and the American Nuclear Society.

LIST OF TABLES

Table		Page

LIST OF FIGURES

Figure		Page

CONTENTS

1. INTRODUCTION

This book addresses the hydrologic features (including water quality and use) and aquatic ecological aspects of the forty-eight contiguous United States. Information on the distribution of endangered aquatic species and on the locations of designated "wild and scenic rivers" is also included. Data are organized by major watersheds — the eighteen water resource regions (WRRs) as defined by the Water Resources Council (see Fig. I). Within regions a degree of internal environmental consistency exists, although most regions exhibit considerable variability. We have attempted to highlight salient environs of each region to provide the reader with an impression of the dominant or otherwise important habitats represented. Referral to other literature is encouraged by extensive citation of publications relating to specific topics.

A comprehensive regional analysis often requires a multifaceted approach that includes investigation of such factors as terrestrial ecology, land use, and political and social structure (Klopatek et al. 1979). However, regions defined on such bases usually do not correspond to watershed units.

The discussion of hydrologic and ecological features is integrated for each water resource region. The separate treatment of endangered and threatened species and their critical habitats and of wild and scenic rivers tabulated by WRR follows.

Fig. I. Water resource regions of the conterminous United States. Source: U.S. Geological Survey (1977).

NEW ENGLAND

Hydrology, Water Quality, and Water Use

The New England region has an area of about 150,000 km^2 (59,000 sq miles) (U.S. Geological Survey 1977) and includes the glaciated Appalachians and adjacent portions of New England. Major rivers, such as the Connecticut, Housatonic, Penobscot, and Kennebec, flow into the Atlantic Ocean or Long Island Sound. In addition to the major reservoirs (Quabbin and Scituate), the area contains numerous natural lakes such as Moosehead and Winnipesaukee. Coastal areas include the mouths of the many seaward-draining streams and large bays such as Narragansett and Boston.

Unpolluted surface waters in the New England region typically are soft (<60 mg/liter hardness as $CaCO_3$). Levels of total dissolved solids (TDS) are <120 mg/liter except for southern Vermont and western Connecticut and Massachusetts, where TDS of up to a few hundred milligrams per liter are found. Total suspended solids (TSS) levels are <270 mg/liter (Geraghty et al. 1973). Both TDS and TSS levels are considered low. Regional water quality problems include increased loads of suspended-solids from erosion, especially as a result of silviculture in New Hampshire and Maine, and excessive nutrient levels in Vermont, Maine, and New Hampshire waters. Toxic pollutants affecting regional surface water quality include heavy metals, pesticides, phenols, and PCBs (polychlorinated biphenyls) (U.S. Environmental Protection Agency 1977b). The New England region also has problems with severe thermal pollution (Geraghty et al. 1973). However, water quality in the Naugatuck and Pemigewasset rivers, which historically have experienced degradation as a result of municipal and industrial discharges, has responded favorably to improved waste treatment programs, and their usefulness for fisheries and recreation is returning (U.S. Environmental Protection Agency 1977b).

Bedrock aquifers in the New England region are generally unproductive; exceptions are the Hudson section of the Valley and Ridge topographic province and the southern Adirondack Mountains. Here, Paleozoic limestones and minor sandstone formations function as aquifers. Fractured granitic and volcanic rocks associated with metamorphics along the mountain ranges yield small amounts of water to wells. Buried and extant stream valley sediments comprised of sand and gravel constitute the most productive groundwater reservoirs in the region. The most areally extensive aquifers are glacial outwash deposits that occur along the coasts of Maine and

3

Massachusetts and in the Connecticut River Valley. Glacial
till and unstratified drift cover much of the region; however,
well yields are low (Walton 1970).

In the New England region, average annual runoff is
equivalent to 250×10^6 m^3/day [67 billion gal/day (Bgd)].
Surface freshwater withdrawals in 1975 were estimated to be
17×10^6 m^3/day (4.4 Bgd); most of the 53×10^6 m^3/day
(14 Bgd) total off-channel water withdrawal (for public
supplies, rural uses, irrigation uses, and self-supplied
industrial and electrical generation uses)* was of saline
water, which accounted for 35×10^6 m^3/day (9.3 Bgd). Use of
groundwater was relatively small, accounting for only
2×10^6 m^3/day (0.6 Bgd). About 9% of the total freshwater
withdrawn was consumed (evaporated or incorporated into
products) (U.S. Geological Survey 1977).

The primary user of water in the region is the self-
supplied industrial sector, including electricity-generating
utilities (Table I). Saline water is used about five times
as much as freshwater for condenser and reactor cooling.
Lesser withdrawals of water are accounted for by public
supplies (which consume the most freshwater of any sector),
rural uses, and irrigation. Generation of hydroelectric power
requires approximately 490×10^6 m^3/day (130 Bgd) of the
national total of 12.5×10^9 m^3/day (3300 Bgd), making it the
eighth-ranking region in this respect (U.S. Geological Survey
1977). Projections to the year 2000 indicate probable water-
supply problems for the region as a result of increased
electrical generation. The Boston, Massachusetts, area is
currently stressed during prolonged drought periods, and
increased demand on New England's water resources is forecast,
particularly in the Connecticut River and Long Island Sound
areas. Increased future water needs may have to be satisfied
by more ocean cooling, expanded use of either freshwater
resources in southcentral Maine or water resource development
or both. The coastal portions of the region (Connecticut,
Rhode Island, Massachusetts, southern Vermont and New Hampshire, and coastal Maine) are among the nation's most critical
areas in terms of water-supply problems that may constrain
energy development; water supplies may be inadequate for power

---

*Off-channel uses are distinquished from in-channel uses,
that is, "uses taking place within the river channel itself
. . . (including) water used for hydroelectric power generation" and nonwithdrawal uses, that is, "water used for navigation, sport, fishing, freshwater discharges into estuarine
areas in order to maintain proper salinity, and the disposi-
tion and dilution of waste water" (U.S. Geological Survey
1977).

Table I. Off-channel water withdrawals (10^6 gpd)[a] in the water resource regions, by sector, 1975[b]

Water resources region	Withdrawals by sector			
	Public supplies (industrial, commercial, domestic)	Rural (domestic, livestock)	Irrigation	Self-supplied industrial and electrical generation
New England	1400 (180)	120 (44)	57 (57)	12,70C (160)
Mid-Atlantic	5300 (760)	470 (180)	230 (200)	46,00C (470)
Tennessee	330 (40)	79 (57)	7.2 (6.9)	10,00C (180)
South Atlantic–Gulf	3100 (930)	750 (560)	3,100 (1,500)	10,00C (750)
Great Lakes	3100 (410)	370 (140)	99 (94)	32,400 (420)
Ohio	2200 (240)	490 (300)	34 (32)	33,020 (640)
Upper Mississippi	3000 (170)	450 (300)	150 (140)	15,015 (190)
Lower Mississippi	750 (310)	130 (120)	4,900 (4,000)	10,300 (1,100)
Arkansas–White–Red	930 (330)	330 (310)	10,000 (8,000)	3,940 (370)
Texas Gulf	1400 (560)	240 (240)	7,100 (6,500)	13,400 (680)
Rio Grande	350 (190)	63 (54)	4,900 (3,200)	129 (75)
Souris–Red–Rainy	48 (20)	40 (27)	42 (41)	230 (6)
Missouri Basin	1200 (290)	620 (550)	28,000 (14,000)	4,736 (120)
Upper Colorado	77 (25)	23 (17)	3,700 (1,500)	255 (87)
Lower Colorado	510 (240)	85 (74)	7,520 (5,700)	430 (240)
Great Basin	380 (140)	76 (25)	6,000 (3,400)	396 (69)
California	3700 (1500)	230 (130)	35,000 (21,000)	11,800 (210)
Pacific Northwest	1200 (230)	310 (230)	28,000 (9,900)	3,441 (310)

[a] 1 million gpd = 3.79 × 10^3 m^3/day.

[b] Data in parentheses represent freshwater consumption.

Source: U.S. Geological Survey 1977.

generation and related cooling needs by 1985 (Water Resources
Council 1974; Peterson and Sonnichsen 1976; Dobson and
Shepherd 1979).

Aquatic Ecology

    Major factors influencing the biotic characteristics of
the aquatic resources in this area are the sparsity of sedimentary rocks and the previous glaciation of the entire area
(Geraghty et al. 1973). Glaciation has resulted in the
occurrence of numerous lakes, ponds, bogs, and streams of
relatively low hardness and low natural productivity (Brooks
and Deevy 1966). Water bodies in southern New England are
generally higher in total hardness because of the occurrence
of limestone bedrock. In addition, their higher natural
nutrient levels and the addition of large quantities of
municipal wastes make them the most productive.
    The bogs that are prevalent throughout much of the region
are very similar to those found in the northern Midwest, in
the alpine zone of the Rockies, and on the Coastal Plain. Bog
waters are typically acid (pH 4-7), very low in buffering
capacity, and high in humic acids (Brooks and Deevy 1966).
Few fish exist in the northern bogs because of shallowness and
low pH, but a great diversity of phytoplankton and zooplankton
often occurs (Macan 1974).
    Freshwater marshes in the region are typical of those
found in the temperate United States. Most are dominated by
cattail (*Typha* spp.) and are highly productive, particularly
in comparison to adjacent lakes (Reid 1961). Because of the
shallowness of these marshes and their tendency to become
seasonally anoxic, few fish are present year-round, although
minnows and bullheads are common in the deeper marshes.
Numerous algal, angiosperm, and invertebrate species exist in
freshwater marsh environments. Because of the quantity of
autotrophic biomass produced, these systems serve as efficient
purifiers of the water that trickles through them; thus, they
are likely to provide some degree of protection against pass-
ing pollutants onto adjacent water bodies (Hynes 1971).
    Most lakes in the region are of glacial origin, although
a few artifical impoundments do exist. The greatest density
of lakes and the largest lakes in terms of area occur in the
north; three have areas greater than 100 km^2 (>39 sq miles)
(Brooks and Deevey 1966). Many of the northern New England
lakes remain relatively pristine, although increased recreation, lumbering, agriculture, and an expanding population have
compromised their quality. The largest lakes are typically
highly oligotrophic and contain little macrophyte development
and few benthic invertebrates. These possess a cold-water
sport fishery dominated by whitefish, trout, and freshwater
salmon (Brooks and Deevey 1966).

Several large rivers occur in the region, but considerable stress from industrial, agricultural, and municipal
wastes has greatly modified the biota of most of these
(Brooks and Deevey 1966). Numerous smaller streams that are
more isolated from perturbations also exist, particularly in
northern New England. These are typically cold and well-
oxygenated and commonly contain stable populations of brook
trout (Brooks and Deevey 1966).

The region contains several estuaries and bays, many of
which display naturally high productivity. However, the
influx of large quantities of heavy metals, pesticides, and
other wastes has greatly altered the biota of most of these
(Reid 1961).

A regional problem that may adversely affect the biota is
the increased acidity of surface waters as a result of acid
rain. This phenomenon most severely affects poorly buffered
waters, such as those found in this region (Hornbeck et al.
1977). Generally, the causal factors are airborne sulfates
and nitrogen oxides — from smelting, fossil fuel combustion,
and industrial processes — that are brought to earth in rain
as acids. Increased fossil fuel combustion, if not coupled
with desulfurization equipment, can be expected to exacerbate
the acid rain problem.

In general, heavy metal toxicities are also greater in
poorly buffered waters than in those with high carbonate-
bicarbonate concentrations (National Academy of Sciences and
National Academy of Engineering 1972). Thus, this water
resource region (WRR) is more susceptible than many to toxic
effects from coal pile and ash pile leachates.

MID-ATLANTIC

Hydrology, Water Quality, and Water Use

The Mid-Atlantic basin has an area of about 264,000 km^2
(102,000 sq miles) (U.S. Geological Survey 1977) and includes
major rivers such as the Hudson, Delaware, Susquehanna,
Potomac, James, and Rappahannock that flow into the Atlantic
Ocean and coastal bays. In northeastern New York—northern
Vermont drainage is northward toward the St. Lawrence River in
Canada. Major reservoirs are generally not important surface-
water features, although some tributaries to the Delaware and
Hudson systems are impounded. The most prominent natural
lakes are found in the northern parts of the region in New
York (Lakes George and Champlain, the latter shared with the
state of Vermont) and Pennsylvania (Lake Wallenpaupack).
Coastal areas include the Chesapeake estuary (drowned-valley)
system (Chesapeake Bay and the mouths of the Potomac, Rappahannock, York, and James rivers), Delaware Bay, the mouth of

the Hudson River (including Upper and Lower New York Bay), and
the extensive bays between the mainland and barrier beaches
from Virginia to Long Island.

Surface waters in the Mid-Atlantic region are of medium
hardness (60 to 120 mg/liter hardness as $CaCO_3$) in the James
River basin, the lower Hudson and Delaware river basins, and
portions of the upper Susquehanna and Potomac river basins.
They are typically soft (<60 mg/liter hardness) elsewhere.
Levels of TDS in surface waters may reach a few hundred
milligrams per liter in the Mohawk-Hudson river system, in
areas of eastern Pennsylvania and northern New Jersey, and in
a belt including the highlands area of central Maryland and
the Virginia-West Virginia border. Isolated areas in eastern
Pennsylvania may have surface waters exceeding 350 mg/liters
TDS. In other parts of the Mid-Atlantic region, TDS levels are
generally below 120 mg/liter. Stream TSS levels, although
generally less than 270 mg/liter throughout the region, may be
higher (270 to 1900 mg/liter) in isolated Piedmont areas of
eastern Pennsylvania, Maryland, and Virginia (Geraghty et al.
1973). Woodruff and Hewlett (1970) mapped the hydrologic
response characteristics (the percent of annual precipitation
that appears as quick flow in small streams), an index of
flood potential, for much of the eastern United States. The
parts of the Mid-Atlantic WRR in their study area generally
showed a stable flow response to precipitation (a hydrologic
response of ≤8%), although some Piedmont areas had a
"flashier" hydrologic response (8-20%).

Regional water quality problems include severe thermal
pollution, pesticides in the lower reaches of major river
basins, and heavy metals, particularly in the more northern
states (New York and Pennsylvania) (Geraghty et al. 1973).
Other toxic pollutants of concern in the region include Kepone
(James River), phenols, and PCBs (U.S. Environmental Protection Agency 1977b). Acid mine drainage in the anthracite coal
area of northeastern Pennsylvania also adversely affects water
quality. The combined acid discharged by the major river
systems in eastern Pennsylvania actually exceeds that of the
bituminous Allegheny-Monongahela river basin of western
Pennsylvania (part of the Ohio Water Resource Region) (U.S.
Environmental Protection Agency 1977b). Increased nutrient
loads and lake eutrophication are problems in Maryland,
Delaware, and New York (U.S. Environmental Protection Agency
1977b).

Major rivers such as the Hudson may deteriorate within a
130-km (80-mile) reach near Albany, New York; other rivers
such as the Mohawk and Susquehanna in New York have improved
since the installation of water treatment plants (U.S. Environmental Protection Agency 1977b). Saltwater encroachment in
coastal areas as a result of groundwater pumping may increase
dissolved solids in groundwater to 3000 mg/liter (U.S.

Environmental Protection Agency 1977b; U.S. Geological Survey 1970).

In the Mid-Atlantic Region, the estimated amount of groundwater in storage ranges from 5.3×10^{11} to 1.3×10^{12} m^3 (4.3×10^8 to 1.1×10^9 acre-ft) (Sinnott and Cushing 1978). Table II shows the estimated groundwater storage and some general aquifer characteristics for the region. The aquifers are of four general types: unconsolidated coastal plain sediments; crystalline metamorphic and igneous rocks; consolidated sedimentary rocks; and glacial sediments (Walton 1970).

The unconsolidated deposits of the Atlantic Coastal Plain consist primarily of sand, clay, and gravel. The sediments range in thickness from 0 to 2,438 m (0 to 8,000 ft) and are hydraulically interconnected to varying degrees. Wells completed in surficial material no deeper than 91 m (300 ft) can produce up to 126 liters/sec (2000 gpm) (Sinnott and Cushing 1978).

The crystalline igneous and metamorphic rocks produce water from fractured and weathered zones. Well yields are typically less than 3 liters/sec (50 gpm); although yields of 0.9 liters/sec (15 gpm) are considered average, some wells produce as much as 25 liters/sec (400 gpm). The groundwater quality is excellent (Sinnott and Cushing 1978).

The consolidated sedimentary rock aquifers are primarily fractured sandstone and solutioned limestone/dolomite. Although some well yields from the folded and faulted rocks of the Valley and Ridge province exceed 63 liters/sec (1000 gpm), most wells produce much less. The groundwater derived from limestone is hard; alkalinity is also a problem. However, the general quality is good to excellent. The sedimentary rocks of the Appalachian Plateaus province are approximately horizontal, and the limestone and sandstone bodies are interbedded with coal seams and shales. The average water well produces from 3.8 to 12.6 liters/sec (60-200 gpm). The groundwater quality is usually good, except in the vicinity of coal-mine areas, where high iron concentrations become a problem (Sinnott and Cushing 1978).

Glacial outwash deposits are a significant source of groundwater in the region. Many wells are capable of producing more than 63 liters/sec (1000 gpm). Groundwater from the glacial deposits is good except locally, where hardness and iron constitute a problem (Sinnott and Cushing 1978).

The aquifers are recharged directly at outcrop areas or by leakage from adjacent aquifers. Discharge is primarily to streams and wells, with some interaquifer flow.

In the Mid-Atlantic Water Resource Region, average annual runoff is equivalent to 320×10^6 m^3/day (84 Bgd). Of the total off-channel water withdrawal of 200×10^6 m^3/day (52 Bgd) in 1975, surface freshwater withdrawals accounted for

Table II. Estimates of groundwater in storage in hydrogeologic units of the Mid-Atlantic region

Hydrogeologic unit	Approximate area [km² (sq miles)]	Average productive thickness [m (ft)]	Average available drawdown [m (ft)]	Average storage coefficient	Estimated storage [m³ (acre-feet)]
Coastal Plain province (sand, clay, gravel)	51,800 (20,000)	76 (250)	46 (150)	0.1–0.2	2.4×10^{11} – 4.7×10^{11} (2.0×10^{8} – 3.8×10^{8})
Piedmont, New England, and Adirondack provinces (crystalline rocks)	77,200 (29,800)	76 (250)	61 (200)	0.001–0.05	4.5×10^{9} – 2.4×10^{11} (3.7×10^{6} – 2.0×10^{8})
Blue Ridge province (crystalline rocks)	5,400 (2,100)	61 (200)	46 (150)	0.001–0.02	2.7×10^{8} – 4.9×10^{9} (2.2×10^{5} – 4.0×10^{6})
Valley and Ridge province (consolidated sedimentary rocks)	74,300 (28,700)	76 (250)	61 (200)	0.05–0.1	2.3×10^{11} – 4.5×10^{11} (1.9×10^{8} – 3.7×10^{8})
Champlain Valley (consolidated sedimentary rocks)	7,000 (2,700)	46 (150)	30 (100)	0.05–0.1	1.1×10^{10} – 2.1×10^{10} (8.9×10^{6} – 1.7×10^{7})
Appalachian Plateaus province (consolidated sedimentary rocks)	64,000 (24,700)	61 (200)	46 (150)	0.02–0.05	5.7×10^{10} – 1.5×10^{11} (4.6×10^{7} – 1.2×10^{8})

Source: Adapted from Sinnott and Cushing (1978).

about 83×10^6 m^3/day (22 Bdg); saline water withdrawals
averaged about 102×10^6 m^3/day (27 Bdg). Groundwater contributed less than 11×10^6 m^3/day (3 Bgd). About 7% of the
freshwater withdrawn was actually consumed (U.S. Geological
Survey 1977).

The primary user of water in the region is the self-
supplied industrial sector, including electricity-generating
utilities (Table II). Almost twice as much saline water as
freshwater is used for condenser and reactor cooling. In
contrast to water withdrawals, the greatest freshwater consumption is accounted for by the public supplies sector.
Generation of hydroelectric power uses approximately
830×10^6 m^3/day (220 Bgd), the Mid-Atlantic Region being the
fifth-ranking region in this respect (U.S. Geological Survey
1977).

Projections to the year 2000 point to probable water-
supply problems. Freshwater supply shortages, currently
experienced in the Delaware and Potomac rivers, will probably
be a factor in the Hudson and Susquehanna river areas as well.
However, coastal siting, which is necessary for saline water
use (currently the practice in the Lower Hudson, Delaware, and
Chesapeake Bay areas), may be limited by site availability.
The Water Resources Council has categorized the area, including southeastern New York, New Jersey, Delaware, and eastern
Pennsylvania, as being among the nation's most critical in
terms of energy-related water-supply problems. Available
water supplies may be inadequate for power generation and
related cooling needs by 1985 (Water Resources Council 1974;
Peterson and Sonnichsen 1976; Dobson and Shepherd 1978).

Aquatic Ecology

The Mid-Atlantic WRR contains a wide variety of physio-
graphic features such as ridge and valley topography, Blue
Ridge mountain terrain, piedmont, coastal plain, and glacial
plains (Strahler and Strahler 1976) that affect the area's
aquatic ecological resources. The upper portion of the
region, including parts of New York, New Jersey, and Pennsylvania, was glaciated and contains numerous lakes, wetlands,
and streams (Berg 1966). Because of edaphic factors, most of
these water bodies (particularly those in the Adirondack and
Catskill mountains) are highly oligotrophic, and their low
nutrient levels severely limit fish production (Berg 1966).
Other lakes in the area are highly productive, both those in
lowlands containing limestone, which contributes to their
usually medium hardness, and those receiving municipal sewage
and fertilizer runoff (Berg 1966). These areas, which include
New York's Finger Lakes, contribute greatly to the four
million acres of fishable freshwaters in this and the New

England WRR (Geraghty et al. 1973). The trophic status of oligotrophic Lake George and eutrophic Canadarago Lake, both in New York, is discussed in U.S. Environmental Protection Agency (1977d). Major fishing resources in the glaciated area include brook and brown trout in the mountain streams and muskellunge and centrarchids in the mesotrophic lowland lakes (Berg 1966).

Unglaciated portions of the region contain very few water bodies and virtually no natural lakes. Hot springs occur in a few areas, but their total water volume is insignificant. From an ecological standpoint, however, they are of great interest as research areas because the biotic communities they harbor are very different from those in adjacent waters (Yount 1966). The coastal plain (particularly of New Jersey) has several bogs similar to those found in the New England WRR (Berg 1966). The region also contains numerous highly productive estuaries — important nursery grounds for many commercial and sport fish and shellfish.

This WRR receives the greatest municipal sewage loading of any region (Geraghty et al. 1973), a condition that has caused severe biotic degradation of its major rivers and of many of its lakes (Berg 1966). Coal mining in Pennsylvania and Virginia has also caused considerable harm by eliminating trout from some areas and by virtually eliminating all biota from others (Berg 1966). Major factors contributing to this degradation include sedimentation, acid drainage, and heavy metals input. Heavy metals from industries, pesticides, and the highest regional output of thermal effluents have contributed to the demise of the major rivers (Geraghty et al. 1973). Although few of these rivers are actively fished and recreation is curtailed on many, some recent progress has been made in reversing their degradation (Berg 1966).

TENNESSEE

Hydrology, Water Quality, and Water Use

The Tennessee River basin has an area of about 106,000 km^2 (41,000 sq miles) (U.S. Geological Survey 1977) and contains the heavily impounded Tennessee River and its tributaries from headwaters in southwestern Virginia and western North Carolina to its confluence with the Ohio River at Paducah, Kentucky. Major impoundments include Norris, Watts Bar, Chickamauga, and Kentucky lakes. Principal tributaries to the Tennessee River include the Holston and French Broad rivers, which join to form the Tennessee River at Knoxville, Tennessee, and downstream tributaries such as the Clinch, Hiwassee, Sequatchie, Elk, and Duck rivers.

Surface waters in the Tennessee region vary from soft
(<60 mg/liter hardness as $CaCO_3$) in western and southcentral
Tennessee to medium (60 to 120 mg/liter) in east Tennessee and
northern Alabama and Georgia. Some headwaters in southwestern
Virginia have harder water (120 to 180 mg/liter). Total
dissolved solids typically range from 120 to 350 mg/liter.
Waters having less than 120 mg/liter TDS drain the western
slopes of the Appalachians along the Tennessee-North Carolina
border. Impoundment of the Tennessee River system reduces TSS
levels in the mainstem, although tributaries have TSS loads in
the 270 to 1900 mg/liter range (Geraghty et al. 1973).

Eutrophication is a problem in some reservoirs that have
long retention times. Industrial and municipal discharge has
contributed to water quality degradation, particularly in
populated and developed areas such as Chattanooga, Kingsport,
and Knoxville, Tennessee. Adverse effects on pH, dissolved
oxygen levels, and suspended and dissolved solids have been
reported (U.S. Environmental Protection Agency 1977b).
Elevated levels of mercury in sediments and biota have been
reported from the Tennessee River in northern Alabama and from
the North Fork of the Holston River (U.S. Environmental Protection Agency 1977b). Water quality problems related to
surface mining of coal have occurred in the Clinch River basin
(U.S. Environmental Protection Agency 1977b). Water quality
in the French Broad River, which had elevated biochemical
oxygen demand (BOD), suspended solids, and heavy metals from
industrial-municipal pollutants, has improved following construction of waste treatment facilities (U.S. Environmental
Protection Agency 1977b). Woodruff and Hewlett (1970) found
that hydrologic response (see section on Mid-Atlantic region)
in the Tennessee WRR varied from stable (≤8%) to "flashy" (up
to 24%).

The Tennessee Region is underlain by various rock types,
several of which are productive groundwater reservoirs. An
estimate 7.31×10^{11} m^3 (2.58×10^{13} ft^3) of groundwater is
available from storage within the region (Zurawski 1978). The
aquifers are of three basic types: unconsolidated sand and
gravel, carbonate bedrock, and noncarbonate bedrock (which
includes crystalline rocks as well as other sedimentary
bedrock).

The Coastal Plain province (Fig. II), an area of about
10,400 km^2 (4,000 sq miles) in the westernmost portion of the
region, is underlain by thick sand, gravel, and clay sequences
of the Mississippi Embayment. Well yields ranging from 12.6
to 37.9 liters/sec (200 to 600 gpm) in this area are not
uncommon. The two principal aquifers are the Coffee Sand and
the McNairy Formation (Sand). Wells completed in the former
characteristically yield up to 18.9 liters/sec (300 gpm);
wells producing from the McNairy Formation can yield from 31.6
to 63.1 liters/sec (500 to 1000 gpm) (Zurawski 1978). Other

13

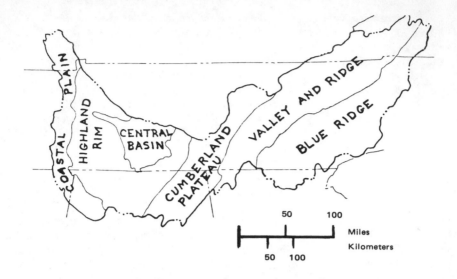

Fig. II.   Provinces within the Tennessee region.

unconsolidated aquifers occur within the region. Stream
alluvium comprises the best aquifer in the unglaciated Appalachians, the easternmost portion of the region (Walton 1970).
Thick regolith (weathered rock) in places forms a productive
aquifer where it overlies solutioned carbonate bedrock (Moran
1977).
        Carbonate rock aquifers occur within about one-half of
the total area of the region. The central portion of the
region (Central Basin, Fig. II) is underlain by a sequence of
approximately horizontal early Paleozoic carbonates (Rima,
Moran, and Woods 1977). Late Paleozoic (Mississippian)
carbonates are the most areally extensive aquifers in the
region, being associated with the Highland Rim and Cumberland
Plateau provinces (Fig. II). Moderately to steeply dipping
carbonate bedrock aquifers occur in the thrust-faulted Valley
and Ridge province (Fig. II) to the east (DeBuchananne and
Richardson 1956; McMaster 1967). The Knox Group (Dolomite) is
the most significant aquifer in this portion of the region,
with well yields of up to 63.1 liters/sec (1000 gpm) (Zurawski
1978). A few isolated areas underlain by carbonates capable
of producing large amounts of groundwater occur in the Blue
Ridge (Unglaciated Appalachians) province (Fig. II) (Zurawski
1978). Well yields vary greatly from one carbonate unit to
another and within any individual formation. Production is
dependent upon both the size and number of solution openings
encountered during drilling and the presence and thickness of
overlying weathered rock material (Table III).

14

Table III. Probable well yields from the
major aquifer types of the
Tennessee region

Aquifer type	Yield per well [liters/sec (gpm)]
Unconsolidated sand	6.3–31.6 (100–500)
Carbonate rocks with regolith	6.3–18.9 (100–300)
Carbonate rocks without regolith	0–18.9 (0–300)
Noncarbonate rocks with regolith	1.6–6.3 (25–100)
Noncarbonate rocks without regolith	Unknown

Source: Adapted from Zurawski (1978).

Areas where noncarbonate aquifers are used are restricted approximately to the eastern one-half of the region. The Cumberland Plateau province (Fig. II) is an area typified by carboniferous sequences of sandstone, conglomerate, shale, and coal, which generally have only fracture permeability. Well yields average less than 3.2 liters/sec (50 gpm). Shales and sandstones of the Valley and Ridge province to the north and east (Fig. II) constitute poor aquifers, whereas the fractured crystalline rocks of the Blue Ridge where fractures and fault zones occur contain groundwater (Zurawski 1978).

The quality of groundwater is variable, but is generally suitable for public drinking-water supplies. Four percent of well and spring analyses show TDS concentrations in excess of 1000 mg/liter (Zurawski 1978). Table IV gives water quality analyses by province.

Approximately 20 to 25% of the total precipitation within the region (9.64×10^5 liters/sec or 1.53×10^7 gpm) is recharged to the groundwater regime (Zurawski 1978). Recharge takes place primarily through losing streams and infiltration of rainfall over an aquifer outcrop area. Discharge is to wells, springs, and gaining streams.

In the Tennessee Water Resource Region average annual runoff is equivalent to 160×10^6 m^3/day (41 Bgd). Of the

Table IV. Groundwater quality of the Tennessee region
(Median values in mg/liter)

Physiographic Province	Silica	Iron	Calcium	Magnesium	Sodium	Potassium	Bicarbonate	Sulfate	Chloride	Fluoride	Nitrate
1 Blue Ridge	16	0.05	4.6	0.5	3.3	0.8	29	1.6	0.9	0.1	0.1
2 Valley and Ridge	10	0.09	38	12	4.5	4.5	178	5	3.5	0.0	3.9
3 Cumberland Plateau	6.4	1.0	16	3.5	3.9	1.2	60	10	2.8	0.1	0.8
4 Highland Rim	11.5	0.00	39	3.8	3.4	3.4	146	4.2	4.0	0.1	1.9
5 Central Basin	7.3	0.08	79	9.7	4.4	1.5	256	26	5.0	0.3	0.5
6 Coastal Plain	14	0.05	16	4.6	22	3.4	95	12	4.7	0.2	0.7

Source: Adapted from Zurawski (1978).

total off-channel water withdrawal of about 42×10^6 m^3/day
(11 Bgd) in 1975, surface freshwater withdrawals accounted for
about 38×10^6 m^3/day (10 Bgd), whereas groundwater withdrawals contributed less than 1×10^6 m^3/day (0.3 Bgd). About
3% of the freshwater withdrawn was actually consumed (U.S.
Geological Survey 1977). The primary user of water (and
consumer of freshwater) in the region is the self-supplied
industrial sector, including electricity-generating utilities
(Table I). Generation of hydroelectric power, a system controlled by the Tennessee Valley Authority, uses approximately
910×10^6 m^3/day (240 Bgd). The Tennessee region ranks below
only the Great Lakes and Pacific Northwest regions in this
respect (U.S. Geological Survey 1977). Projections to the
year 2000 indicate no major energy-related, low-flow, water-
supply problems, primarily because of the Tennessee Valley
Authority's ability to regulate streamflow (Dobson and Shepherd
1979).

Aquatic Ecology

    This WRR generally has steep topography, and none of it
has been glaciated (Geraghty et al. 1973). Hence, few natural
lakes or wetlands exist in the area; the water resources are
dominated by streams, the larger of which have been extensively impounded (Patterson 1970). The waters have low
productivity, and total dissolved solids and hardness concentrations range from low to moderate.
    Mountain streams in the area are typically clear, cold,
and shallow and harbor a diverse invertebrate assemblage
(Hynes 1972). Red algae have been reported from these
habitats (Parker 1976), and rainbow, brown, and brook trout
are locally abundant (Gerking 1966). Many of the higher-
altitude streams remain relatively pristine, but several have
been extensively polluted from acid mine drainage (Gerking
1966). This disturbance has resulted in simplification of the
food chain, with predominance by the few species that are
tolerant of low pH, high dissolved solids, high turbidity, and
increased levels of metallic and organic toxicants (Hynes
1972).
    The larger rivers have been greatly altered from their
original condition. The construction of many dams has
stabilized stream flows, which formerly varied considerably
during the year, and has created a long series of artificial
lakes where flowing water formerly existed. The effect on the
biota has been profound. For example, plankton are more
numerous, and species of benthic invertebrates and fish have
changed (Gerking 1966). These reservoirs provide an active
warmwater sport fishery dominated by bass, sunfish, crappie,
and walleye, and tailrace waters provide habitat for trout in

some areas (Gerking 1966). In this WRR, 377,000 ha
(930,000 acres) of freshwater exist for fishing (Geraghty
et al. 1973).

Major effluent problems that have affected the biota of
the larger streams include municipal waste, which causes
nuisance plant growth, and heavy metals from industrial
facilities (Geraghty et al. 1973).

SOUTH ATLANTIC-GULF

Hydrology, Water Quality, and Water Use

The South Atlantic-Gulf region has an area of about
700,000 km^2 (270,000 sq miles) (U.S. Geological Survey 1977)
and contains 24 major river systems that drain the southern
Appalachian Highlands southeastward and southward toward the
Atlantic Ocean and the Gulf of Mexico. It also contains
smaller coastal river systems, including those of Florida.
Prominent river systems include the Savannah, Roanoke,
Chattahoochee, Apalachicola, Alabama, and Tombigbee. Major
reservoirs include the John H. Kerr Reservoir, discussed in
U.S. Environmental Protection Agency (1977d), Clark Hill and
Hartwell reservoirs, and lakes Marion, Moultrie, Norman,
Lanier, and Martin. Large natural lakes are not typical
features except in central and southern Florida (Okeechobee is
the second largest natural freshwater lake in the conterminous
United States, excluding the Great Lakes) (Water Information
Center 1970); the Carolina Bays (small lakes of the southeastern Atlantic coast) are of particular interest because of
their obscure geologic origin (Yount 1966). The region has an
extensive oceanic shoreline, and bays are prominent at the
mouth of the many rivers and behind barrier beaches.

Surface waters in the South Atlantic-Gulf region are
generally soft (<60 mg/liter hardness as $CaCO_3$) except for
moderately hard (60 to 120 mg/liter) waters in northern
Alabama and coastal areas in peninsular Florida. The Palm
Beach, Florida, area has even slightly harder water (up to
180 mg/liter). Total dissolved solids (TDS) levels are
generally less than 120 mg/liter except for the Black Warrior-
Lower Tombigbee waters (up to 350 mg/liter TDS) and surface
waters in peninsular Florida (up to and exceeding 350 mg/liter
TDS). Concentrations of TSS are generally low in coastal
areas (<270 mg/liter), with higher levels inland (up to
1900 mg/liter) (Geraghty et al. 1973). Beck et al. (1974)
discuss the geochemistry of some rivers in the Georgia coastal
plain. These rivers are dominated by organic constituents and
exhibit low pH.

Surface water quality problems in the region have resulted
from sediment runoff from silviculture and mining, nutrient

18

loading of surface waters, and municipal and industrial discharges (U.S. Environmental Protection Agency 1977b). Contamination of aquatic habitats with pesticides is a problem: use of DDT in the region has been particularly high; dieldrin and lindane have been found in surface waters (Geraghty et al. 1973). The Black Warrior drainage in Alabama has experienced acidification problems from coal mining operations. However, many surface waters in Georgia and Mississippi are naturally acidic as a result of high concentrations of organic acids (U.S. Environmental Protection Agency 1977b). Trace element contamination of surface waters has also occurred; elevated arsenic levels have been found in the Cape Fear (North Carolina) and Catawba (North and South Carolina) river drainages, and elevated cadmium concentrations have been found in interior and coastal Mississippi and Alabama. Industrial mercury pollution has contaminated the lower Tombigbee-Alabama river basin, the lower Savannah River, and waters of coastal Georgia (Geraghty et al. 1973). Saltwater intrusion into groundwater aquifers has resulted from pumping, especially in Florida. In Florida, the canal systems associated with land development reduce the area of wetlands. Construction and operation of waste treatment facilities have resulted in improvement of water quality in many streams, such as the Pearl River, and lower reaches of the Savannah and Tombigbee rivers (U.S. Environmental Protection Agency 1977b). Woodruff and Hewlett (1970) found that the hydrologic response (see section on Mid-Atlantic region) of the South Atlantic-Gulf WRR varied from stable (<4%) to "flashy" (up to 24%); most areas were more moderate, with 4-12% of annual precipitation in small watersheds appearing as quick-response streamflow.

Total groundwater in storage with TDS concentration of less than 3000 mg/liter is estimated at 1.7×10^{13} m^3 $(1.4 \times 10^{10}$ acre-ft). The average total groundwater availability is conservatively estimated at 3.4×10^6 liters/sec $(5.4 \times 10^7$ gpm) (Cederstrom et al. 1979). The principal aquifers consist of highly permeable clastics and limestones of the Coastal Plain.

Aquifers in deposits of Cretaceous age have an areal extent of at least 112,000 km^2 (70,000 sq miles). Well yields as great as 694 liters/sec (11,000 gpm) have been reported; however, maximum expected yields from the Cretaceous aquifers are more on the order of 158 liters/sec (2500 gpm). The TDS concentration is less than 1,000 mg/liter (Cederstrom et al. 1979).

The most productive and utilized source of groundwater in the region is the lower Tertiary limestone aquifer (in Florida, the Floridan aquifer). The aquifer, which underlies an area of approximately 145,000 km^2 (90,000 sq miles), ranges in thickness from 15 to 762 m (50 to more than 2500 ft). The aquifer contains potable water to depths of 610 m (2000 ft);

19

the water, however, is generally hard — at least 100 mg/liter.
Recharge is generally from precipitation over outcrop areas.
Discharge is to wells, springs, and streams. Wells withdraw
millions of liters per day; Silver Springs and Rainbow Springs
alone discharge 3.8×10^9 liters (1×10^9 gallons) per day
(Cederstrom et al. 1979).

In the South Atlantic–Gulf region, average annual runoff
is equivalent to 747×10^6 m^3/day (197 Bgd). Of the total
off-channel water withdrawal of about 160×10^6 m^3/day
(43 Bgd) in 1975, surface freshwater withdrawals accounted for
about 91×10^6 m^3/day (24 Bgd), groundwater withdrawals about
21×10^6 m^3/day (5.5 Bgd), and saline surface water about
53×10^6 m^3/day (14 Bgd). About 13% of the freshwater withdrawn was actually consumed (U.S. Geological Survey 1977).
The geographic distribution of water use in the region is
crucial. Although the total runoff is the largest of all the
eastern regions, many of the population centers (such as
Birmingham, Alabama; Atlanta, Georgia; and Charlotte, Green-
ville, and Winston-Salem, North Carolina) are in inland headwater areas where low-flow problems result from periodic
droughts. Important river mainstems, such as those of the
Alabama, Tombigbee, and Apalachicola, pass through less
populated areas. This disparity is also great in southern
Florida, which has large population centers (Miami and Palm
Beach) and high agricultural water use, but no major rivers.
Rainfall in southern Florida is both highly seasonal and
variable from year to year. Because of these geographic and
flow-variation problems, it has been estimated that only about
190×10^6 m^3/day (50 Bgd) of the runoff is currently available
(Water Resources Council 1974; Peterson and Sonnichsen 1976).

The self-supplied industrial and electricity generation
utility sector is the largest user of water in the region,
although irrigation is the largest consumer of freshwater
(Table I). Use of saline water for condenser and reactor
cooling is almost equal to that of freshwater. Generation of
hydroelectric power in the region requires about 800×10^6
m^3/day (210 Bgd), making the region the sixth-ranking in this
respect (U.S. Geological Survey 1977). Water supply projections for the region indicate that, although supplies are
available along much of the coast, increased inland use will
require water resource development and increased use in
southern Florida will necessitate even greater reliance on
saltwater withdrawals (Water Resources Council 1974; Peterson
and Sonnichsen 1976). Peninsular Florida has been rated as
one of the nation's most critical energy-related water-supply
problem areas by the Water Resources Council. By 1985, available water supply may be inadequate for power generation and
cooling needs (Water Resources Council 1974).

This WRR encompasses much of the Eastern Coastal Plain and Piedmont areas. Most of the southern part of the region is characterized by very soft water, whereas Florida (particularly the southern half) generally contains moderately hard to hard waters. Natural productivity of these waters is roughly proportional to hardness (Yount 1966).

Rivers in the northern portion of the region typically arise as mountain streams that become warmer, slower, and more turbid in the Piedmont (Yount 1966). After crossing the fall line, these rivers enter the Coastal Plain and become broad, slowly flowing streams with lower silt content. Many of them terminate in swamps and marshes near the coast (Yount 1966). The natural biota of these rivers changes markedly along their length; that is, trout are prevalent in the undisturbed headwaters whereas only species capable of tolerating very high temperatures [up to 35°C (95°F)] occur in the marshy areas near the rivers' mouths (e.g., cyprinids, centrarchids) (Hynes 1972).

Aside from several reservoirs, the only major lakes in the north are the Carolina Bays. These are small lakes that have peat and sand bottoms, low pHs, and low productivity, but diverse fish communities (Yount 1966). In contrast, the solution lakes of Florida (of which there are thousands) are typically highly productive and alkaline and often contain dense growths of duckweed (*Lemna*), water hyacinth (*Eichornia*), water lettuce (*Pistia*), and water fern (*Salvinia*). The smaller ones are typically anoxic a short distance below the surface, and the only fish present are minnows (particularly *Gambusia affinis*) (Yount 1966). An extensive discussion of the trophic status of one Florida lake (Lake Weir) may be found in U.S. Environmental Protection Agency (1977d). Florida's large lakes harbor a good sport fishery dominated by largemouth bass; the entire region claims the greatest area of fishable freshwaters of any WRR [1.89×10^6 ha (4.68×10^6 acres)] (Geraghty et al. 1973).

This WRR has the greatest areal extent of freshwater swamps and marshes of anywhere in the country (Geraghty et al. 1973). Additionally, much of its coastline consists of salt marshes and mangrove swamps. The inland swamps are typically devoid of large fish but contain many amphibians, invertebrates, algae, and angiosperms. Their water is typically highly colored from fulvic and humic substances and is acidic (Yount 1966). The freshwater marshes are highly productive and usually contain clear, alkaline water; they do not harbor large fish but contain a diversity of other aquatic biota (Reid 1961). Grasses, sedges, reeds, and rushes are the emergent macrophytes of these areas, and the community is very different from that found in more temperate locations of the country (Fassett 1957).

A habitat common to portions of this region is the artesian spring. These springs occur primarily in the karst area of Florida and represent some of the most productive areas in the country (Yount 1966). Their waters are nearly homothermal year-round and are generally hard. In addition to large standing crops of fish, these springs contain large quantities of angiosperms (chiefly *Sagittaria lorata*) and algae (Yount 1966).

Major perturbants of ecosystems in the region include high pesticide inputs, municipal waste effluents, and turbidity from logging and agriculture. Consequently, the region is one of the highest in number of fish kills (Geraghty et al. 1973). Turbidity increases have been particularly destructive in the north, where suspended solids levels of 1200 mg/liter and greater are not uncommon (Yount 1966). Siltation is frequently implicated in the destruction of bottom organisms — by creating food shortages for fish — and in the direct destruction of fish (by gill clogging, covering of spawning grounds, etc.) (Hynes 1971).

GREAT LAKES

Hydrology, Water Quality, and Water Use

The Great Lakes region has an area of about 325,000 km^2 (126,000 sq miles) (U.S. Geological Survey 1977) and includes all of Michigan and portions of eight other states within approximately 161 km (100 miles) of the Great Lakes. In addition to the Great Lakes (Superior, Michigan, Huron, Erie, and Ontario), the streams flowing into them (such as the Kalamazoo-, Menominee, and Maumee rivers), and the St. Lawrence River, which is the outlet of the Great Lakes, the region includes the Finger Lakes area of central New York and Lake Winnebago.

The hardness of surface waters in the Great Lakes Water Resource Region varies locally from soft (<60 mg/liter hardness as $CaCO_3$) to hard (>120 mg/liter). The lowest levels are generally found in those parts of Minnesota, Wisconsin, and the Upper Peninsula of Michigan that border on Lake Superior. Surface waters of moderate (60 to 120 mg/liter) hardness are typically found in northern parts of Wisconsin, Michigan, and in western New York. Hard waters are present in most of the remaining portions of the region, with the hardest waters (180 to >240 mg/liter) in northwestern Ohio and southeastern Michigan. The hardness of the Great Lakes ranges from soft (Superior) to moderate (northern Huron and Michigan and eastern Ontario) to hard (Erie, southern Michigan and Huron, and western Ontario). Total dissolved solids (TDS) concentrations in surface waters of the Great Lakes region range from less than 120 mg/liter in areas bordering the western shores of Lake Superior and that part of New York near the outlet of Lake Ontario, to greater than 350 mg/liter in central Michigan, along the southwestern shores of Lakes Michigan and Erie, and along the southeastern shore of Lake Ontario. Other areas in the region are characterized by intermediate TDS concentrations (120 to 350 mg/liter). Levels of total suspended solids are generally low (<270 mg/liter) except for the Maumee basin area of northwestern Ohio, where levels range from 270 to 1900 mg/liter (Geraghty et al. 1973).

With the exception of nutrient enriched portions of Lake Erie, the open waters of the Great Lakes are generally of good quality. Inshore, harbor, and developed areas suffer from a variety of water quality problems, including excessive nutrient inputs, pesticides, bacterial contamination, erosion (both natural and aggravated), heavy metals, dissolved solids, phenols, PCBs, other organics, and asbestos (U.S. Environmental Protection Agency 1977b). Nutrient loading for Lake Michigan is discussed in U.S. Environmental Protection Agency (1977d). The Detroit and Cuyahoga rivers have been heavily impacted by industrial and municipal effluents; the Detroit River has responded favorably to waste treatment programs, but the assimilative capacity of the lowest reaches of the Cuyahoga River may continue to be overstressed even though waste treatment has been fully instituted (U.S. Environmental Protection Agency 1977b). The Great Lakes region has been characterized as having severe thermal-pollution problems comparable to those of the New England and Mid-Atlantic regions (Geraghty et al. 1973).

An estimated 9.9×10^{11} m^3 (3.5×10^{13} ft^3) of potable goundwater is available from storage in the Great Lakes region (Table V) (Weist, Jr. 1978). The aquifers are of three main types: crystalline bedrock of the Canadian Shield, consolidated sedimentary rocks, and unconsolidated glacial material. Intrusive igneous and metamorphic rocks in the Lake Superior area yield small volumes to wells from fractures. Sandstones and some carbonates are significant aquifers. The sandstone bedrock of Wisconsin, northernmost Illinois, and Michigan produces water from primary porosity zones. Carbonates (limestone and dolomite) produce from permeable reef structures and secondary fractures (Walton 1970).

Bedrock wells generally have adequate yields. Wells completed in sandstones along the western and northern part of the Lake Michigan basin can produce up to 82 liters/sec (1300 gpm). Limestone aquifers are capable of sustaining well yields of up to 63.1 liters/sec (1000 gpm). Crystalline rock and shale units may not yield as much as 0.32 liters/sec (5 gpm), and constitute the poorest aquifers in the region (Weist, Jr. 1978).

23

Table V. Groundwater occurrence in the Great Lakes region

Aquifer system	Yields of highest-capacity wells liters/sec (gpm)	Principal water-bearing units	Areas of greatest potential
		Unconsolidated deposits	
Glacial lake deposits	0.6-1.3 (10-20)	Sandy zones within the silt and clay	
Outwash sand and gravel	3.2-315 (50-5,000)	Highest yields are obtained in areas with good sources of recharge	Indiana — South Bend area Michigan — Jackson, Kalamazoo, area; Au sable, Manitee, and Muskegon River valley New York — Genesee River basin Ohio — northeastern Pennsylvania
Till	0.6-1.3 (10-20)	Sand and gravel lenses	
		Bedrock	
Pennsylvanian	3.2-44 (50-700)	Sandstone and shale — brine and sulfates frequently encountered in lower part	Southern Michigan
Mississippian	3.2-114 (50-1,800)	Sandstone and shale — may contain oil, gas, and brine	Southern Michigan, northeastern Ohio, Pennsylvania
Devonian	3.2-44 (50-700)	Mostly carbonate rocks with some sandstone — may contain oil, gas, and brine	Ohio, southern Michigan, northeastern Indiana
Devonian-Silurian	3.2-32 (50-500)	Carbonate rocks	North Michigan

Table V (continued)

Aquifer system	Yields of highest-capacity wells (gpm) liters/sec	Principal water-bearing units Unconsolidated deposits	Areas of greatest potential
Silurian	3.2–63 (50–1,000)	Mostly carbonate rocks — water is saline in places	Eastern Wisconsin, Illinois, northeastern Indiana, northwestern Ohio
Ordovician	3.2–32 (50–500)	Carbonate rocks and sandstone saline water at depth	Western New York
Cambrian-Ordovician	3.2–82 (50–1,300)	Carbonate rocks and sandstone saline water at depth	Illinois, eastern Wisconsin
Cambrian	3.2–32 (50–500)	Sandstone — saline water at depth	Illinois
Cambrian-Precambrian	3.2–32 (50–500)	Sandstone	North Michigan
Precambrian	6.3–16 (100–250)	Sandstone and crystalline rocks	Minnesota — Mesabi Range

Source: Adapted from Weist, Jr. (1978).

25

Glacial drift varies from several meters to perhaps a hundred meters thick across the region. Much of the glacial material is relatively impermeable till; however, some sand and gravel deposits associated with till are moderately productive. Glacial outwash deposits composed of sand and gravel are locally excellent aquifers (Walton 1970); Weist, Jr. (1978) indicates aquifer yields of up to 315 liters/sec (5000 gpm).

The groundwater within most of the region is hard (more than 120 mg/liter). At depths of less than 152 m (500 ft) the groundwater becomes slightly saline (1000-3000 mg/liter); mineralization increases with depth (Weist, Jr. 1978).

Groundwater recharge takes place through infiltration of snowmelt and rainfall and is induced from surface water by pumpage. As a result of heavy use at major pumping centers like Chicago, Milwaukee, and Green Bay, water levels have declined by as much as 100 m. Discharge points are wells, springs, and streams. The minimum estimated daily groundwater discharge in the region is 1.1×10^6 liters/sec (1.8×10^7 gpm) (Weist, Jr. 1978).

In the Great Lakes Water Resource Region, average annual runoff is equivalent to 280×10^6 m^3/day (75 Bgd). Of the total off-channel water withdrawal of about 140×10^6 m^3/day (36 Bgd) in 1975, surface freshwater withdrawals accounted for about 130×10^6 m^3/day (35 Bgd); groundwater contributed about 6×10^6 m^3/day (1.6 Bgd), one-fourth of which was saline. About 3% of the freshwater withdrawn was actually consumed (U.S. Geological Survey 1977).

The primary user of water in the region is the self-supplied industrial sector, including electricity-generating utilities (Table I). Freshwater consumption by this sector was almost equal to that of the public supplies sector. The saline groundwater withdrawals were by industry; fresh groundwater was withdrawn by all sectors. Generation of hydroelectric power uses approximately 1100×10^6 m^3/day (290 Bgd), with the Great Lakes region second only to the Pacific Northwest region in this respect (U.S. Geological Survey 1977).

A projection of future water availability in this region indicated that the functioning of the Great Lakes as a large reservoir system would result in general availability of water for energy technologies, with no serious low-flow problems. The lakes themselves have dampened seasonal fluctuations, whereas the outflow through the St. Lawrence River is exceeded in the United States only by the Mississippi and Columbia rivers and is stable throughout the year. However, large consumption rates might lower lake levels, aggravating the water quality problems in the populated industrialized centers (Pierce and Vogt 1953; Dobson and Shepherd 1979).

The dominant water resource in this region is the Great Lakes. These vast bodies of water are of low to moderate hardness and have very low to moderate natural productivity (Beeton and Chandler 1966). Typical deepwater invertebrate fauna of the lakes include *Pontoporeia affinis* (an amphipod), *Mysis relicta* (oppossum shrimp), oligochaetes (largely *Limnodrilus*), and the molluscs *Pisidium* and *Sphaerium* (Beeton and Chandler 1966; Ricker 1959). The shallow-water benthic fauna is much like that of adjacent smaller bodies of water of the same productivity; the average density of organisms is roughly $1200/m^2$ for Lakes Huron and Michigan (Beeton and Chandler 1966). Although diatoms are the dominant phytoplankton, even in the highly eutrophic portions of Lake Erie, Lake Erie and Lake Ontario sometimes contain extensive populations of blue-green and green algae (Hutchinson 1967). Copepods are the most common zooplankters, followed by cladocerans (Beeton and Chandler 1966).

Salmonids, which dominated the early sport and commercial fisheries of the Great Lakes, still account for most of the sport-fishing enthusiasm (Beeton and Chandler 1966). Lake trout and the coregonids (whitefish, lake herring, chubs) have been the most important commercial species in the upper lakes, whereas shallow waters and the lower lakes have largely produced perch and catfish (Smith 1972). The fish fauna has changed greatly in the past century because of the following major anthropogenic factors (Smith 1972): (1) intensive selective fishing pressure, (2) invasion by marine species and successful establishment of stocked species, (3) modification of the drainage systems entering the lakes, and (4) progressive physicochemical changes. Because of these influences, production rates of the oligotrophic, cold-water fishes have fallen, and mesotrophic species (e.g., percids) are much more prevalent. Exotic species are also common and, in many areas, constitute a large percentage of the total fish standing crop (Christie 1974). Carp and smelt are introduced fish that have some commercial value (Beeton and Chandler 1966). The alewife and the sea lamprey are species that gained entrance into Lakes Erie, Huron, Michigan, and Superior after completion of the canal bypassing Niagara Falls. Both species are regarded as nuisance organisms, although some effort is being made to commercially harvest the alewife (Beeton and Chandler 1966); control efforts have largely eliminated sea lamprey problems (Christie et al. 1972). Three sport fish that have been recently introduced are apparently well established in the more pristine areas of the lakes. Rainbow trout, coho salmon, and chinook salmon are actively sought by anglers, and the latter two have been successful in cropping some of the alewife production in Lake Michigan (Stauffer 1976; Christie et al. 1972; Christie 1974).

The Great Lakes have incurred considerable alteration by virtue of the high human population density and industrial activity nearby. Thermal inputs are greater than those in nearly any other region, toxicant additions are very heavy locally, and the addition of nutrients via sewage and other high-BOD wastes has accelerated eutrophication considerably (Geraghty et al. 1973). Although Lake Superior has escaped much of the change that has occurred in the other lakes, it too has exhibited biotic changes as a result of man's activities. Interestingly, not all of the changes can be easily categorized as undesirable because increased eutrophication has increased fish production and harvests in some areas (Beeton and Chandler 1966).

Other water bodies within this region include bog lakes (much like those of New England), soft-water oligotrophic lakes, hard-water eutrophic lakes, large turbid streams, clear trout streams, and marshes (Strahler and Strahler 1976). In general, those lakes found in the southern portion of the region (southern Wisconsin, southern Michigan, northern Indiana, and northern Ohio) are eutrophic as a result of the addition of municipal wastes, feedlot wastes, etc. Although several of these support dense stands of macrophytes (Swindale and Curtis 1957) and have engendered little sport-fishing interest, many are highly productive and heavily fished (Knight, Ball, and Hooper 1962; Tanner 1960). The lakes in the northern part of the WRR are largely undisturbed and are similar to those found in the northeast portion of the Souris-Red-Rainy WRR (Tanner 1960). This glacially modified area contains numerous oligotrophic waters (it has one of the highest concentrations of lakes in the world) (Fassett 1930). Aside from the tremendous ecological value of this area's biotic resources, their major use is recreation-related (Johnson and Hasler 1954).

OHIO

Hydrology, Water Quality, and Water Use

The Ohio Water Resource Region, with an area of about 420,000 km^2 (163,000 sq miles) (U.S. Geological Survey 1977), includes the Ohio River and tributary systems (such as the Allegheny, Monongahela, Cumberland, Kanawha, Kentucky, Green, and Wabash rivers). Although the region includes scattered small natural lakes and moderate-sized reservoirs, major reservoirs are limited to the Cumberland River (Barkley, Center Hill, and Dale Hollow reservoirs and Lake Cumberland). Surface waters of the Ohio Water Resource Region range from moderately hard (60 to 120 mg/liter hardness as $CaCO_3$) to hard (>120 mg/liter). The softer waters are generally found

28

in the southern and eastern portions of the region. Surface-
water hardness in parts of Illinois, Indiana, and Ohio falls
in the 180 to 240 mg/liter range. Total dissolved solids
levels also reflect this distribution, with lowest values
(<120 to 350 mg/liter) found in the southern and eastern
portions of the region, and the highest levels (>350 mg/liter)
in parts of Illinois, Indiana, Ohio, and western Pennsylvania.
Total suspended solids (TSS) levels generally vary from less
than 270 mg/liter to almost 2000 mg/liter. The southern
Illinois area has the highest TSS levels, exceeding
1900 mg/liter (Geraghty et al. 1973).

Water quality in the Ohio basin has suffered from both
point-source and nonpoint-source pollutants. Agriculture
(contributing pesticides, nutrients, and sediments), construction, urban runoff, stream channelization, and industrial and
municipal discharges have degraded water quality in terms of
bacterial contamination, nutrients, heavy metals, and other
toxicants (U.S. Environmental Protection Agency 1977b). The
region has been categorized as having major thermal pollution
problems, although not as severe as those of the New England
and Mid-Atlantic regions (Geraghty et al. 1973). The Ohio
region has been particularly affected by coal mining drainage,
with abandoned mines and the diffuse nature of the discharge
exacerbating the situation (U.S. Environmental Protection
Agency 1977b). The Environmental Protection Agency (EPA) has
identified Pennsylvania, Maryland, northern West Virginia,
Ohio, western Kentucky, and the Illinois-Indiana border area
as being affected by acid/ferruginous drainage (U.S. Environmental Protection Agency 1977a). Coal mining in eastern
Kentucky and the Cumberland region in Tennessee also results
in acid mine drainage. Streams in mining areas of eastern
Kentucky have suffered from elevated levels of dissolved
solids, hardness, sulfate, acidity, and chloride. Even where
neutralized mine wastes have been discharged, increased dissolved solids and sulfate levels have been reported. Although
identifiable point-source discharges such as industrial and
municipal wastes have begun to be controlled with waste
treatment facilities (with observed recovery of water quality),
mine drainage has been identified as less amenable to control
with "best available treatment" technologies (U.S. Environmental Protection Agency 1977b). Portions of the Ohio WRR
within the study area of Woodruff and Hewlett (1970) showed a
small-watershed hydrologic response (see section on the Mid-
Atlantic region) ranging from stable (<4%) to "flashy" (>24%).
The watersheds with the most rapid flow response to precipitation were found in northern Kentucky.

Approximately 4.09×10^{11} m^3 ($108,000 \times 10^9$ gal) of
potable groundwater is available from storage in the aquifers
of the Ohio region. Consolidated rock and minor unconsolidated aquifers contain 3.2×10^{11} m^3 ($85,000 \times 10^9$ gal) of

the total amount; outwash and alluvial aquifers may store
as much as 8.7×10^{10} m^3 (23,000 $\times$ 10^9 gal) of potable groundwater (Bloyd 1974).

The groundwater reservoirs in the Ohio region are of four
general types: alluvium, outwash, glaciofluvial deposits, and
bedrock. Alluvium of recent age, consisting of silt, sand,
and gravel, occurs in the valleys of the larger streams. Sand
and gravel glacial outwash deposits are found in the glaciated
valleys of the main tributaries. Glaciofluvial deposits are
comprised of both Pleistocene outwash and recent alluvium;
these occur within the Ohio River valley as well as along its
major tributaries. Well yields from the unconsolidated
aquifers can exceed 31 liters/sec (500 gpm). Thick bedrock
units varying in age from Precambrian to Tertiary underlie the
region. The greatest thicknesses are coincident with the
Appalachian structural basin in the east and the Illinois
structural basin in the west. Maximum well yields of consolidated aquifers range from 6 to 31 liters/sec (100 to
500 gpm) (Bloyd 1974). Characteristics of the significant
aquifers in the Ohio region are given in Table VI.

Groundwater quality within the region varies primarily
with depth. Shallow groundwater generally contains less than
1000 mg/liter TDS. At depths greater than 150 m (500 ft),
however, TDS concentrations may exceed 35,000 mg/liter (Bloyd
1974).

Recharge occurs primarily by infiltration of precipitation. Some seepage from streams takes place. The average
annual regional groundwater recharge is about 1.3×10^8 m^3/day
(35,000 Mgd), which may be as much as 15% of the total annual
precipitation (Bloyd 1974). Discharge takes place through
seepage to streams, pumpage, and underflow to adjacent areas.

The base-year (1960) use of groundwater for municipalities
and rural demands was approximately 3.79×10^6 m^3/day
(1000 Mgd), or about 3% of annual recharge. Industrial use of
groundwater was also equivalent to 3% of annual recharge.
Table VII shows the projected use and recharge of groundwater
by subbasin for the year 2020 (Bloyd 1974).

In the Ohio Water Resource Region, average annual runoff
is equivalent to 474×10^6 m^3/day (125 Bgd). Of the total
off-channel water withdrawal of about 140×10^6 m^3/day
(36 Bgd) in 1975, surface freshwater withdrawals accounted for
about 130×10^6 m^3/day (34 Bgd); groundwater contributed about
7.2×10^6 m^3/day (1.9 Bgd). About 3% of freshwater withdrawn
was actually consumed. Primary groundwater users were the
industrial and public supplies sectors (U.S. Geological Survey
1977). The greatest user of water and consumer of freshwater
in the region is the self-supplied industrial sector, including electricity-generating utilities (Table I). Generation of
hydroelectric power uses approximately 870×10^6 m^3/day

30

Table VI. Characteristics of the consolidated and unconsolidated aquifers in the Ohio region and ranking of aquifers in order of decreasing transmissivity

Aquifer	Hydraulic conductivity [liters/day m² (gpd/ft²)]	Storage coefficient or specific yield
Mad River alluvial aquifer	1410-1580 (4000-4500)	0.25
Ohio River valley outwash and alluvial aquifer	141-2960 (400-8400)	0.5-0.20
Miami River, Scioto River, Upper Muskingum River, and Whitewater River alluvial aquifers	880-1060 (2500-3000)	0.15-0.20
Allegheny, Lower Wabash, and White River alluvial aquifers	704 (2000)	0.15-0.20
Hocking River, Lower Muskingum River, and Upper Wabash River alluvial aquifers	528-704 (1500-2000)	0.15-0.20
Beaver River alluvial aquifer and alluvial aquifers in the minor tributaries north of the Ohio River	176-352 (500-1000)	0.15
Alluvial aquifers in the major tributaries south of the Ohio River	176 (500)	0.10
Mississippian bedrock aquifer (Green River Basin)	>7 (>20)	0.01-0.05
Mississippian bedrock aquifer with glacial cover	>7 (>20)	0.01-0.05
Pennsylvanian bedrock (Allegheny and Pottsville Formations)	>7 (>20)	0.01-0.05
Pennsylvanian bedrock with glacial cover	>7 (>20)	0.01-0.05
Silurian bedrock with glacial cover	>7 (>20)	0.01-0.05

Table VI (continued)

Aquifer	Hydraulic conductivity [liters/day m^2 (gpd/ft^2)]		Storage coefficient or specific yield
Pennsylvanian bedrock (Conemaugh Formation)	>7	(>20)	0.01-0.05
Pennsylvanian and Permian bedrock (Dunkard Group)	>7	(>20)	0.01-0.05
Ordovician bedrock with glacial cover	>7	(>20)	0.01-0.05
Silurian bedrock	>7	(>20)	0.01-0.05
Mississippian bedrock	>7	(>20)	0.01-0.05
Pennsylvanian bedrock (Monongahela Formation)	>7	(>20)	0.01-0.05
Devonian bedrock with glacial cover	>7	(>20)	0.01-0.05
Ordovician and undifferentiated Paleozoic rocks	>7	(>20)	0.01-0.05
Devonian bedrock	>7	(>20)	0.01-0.05

Source: Adapted from Bloyd (1974).

Table VII. Comparison of projected groundwater withdrawals with estimated groundwater recharge for the year 2020

Subbasin	Total[a] withdrawal [m^3/sec (Mgd)]		Estimated groundwater recharge [m^3/sec (Mgd)]	
Allegheny	8.8	(200)	180	(4,100)
Monongahela	8.3	(190)	127	(2,900)
Upper Ohio River	12.7	(290)	70.1	(1,600)
Muskingum	9.2	(210)	70.1	(1,600)
Kanawha-Little Kanawha	14.0	(320)	223	(5,100)
Scioto	6.6	(150)	26.3	(600)
Big and Little Sandy-Guyandotte	4.4	(100)	57.0	(1,300)
Great and Little Miami	29.8	(680)	70.1	(1,600)
Licking			320	(4,800)
White	34.6	(790)		(2,500)
Green–Salt–Lower Ohio River	1.8	(40)	145	(3,300)
Cumberland	4.4	(100)	206	(4,700)
Basinwide industrial use	40.7	(930)		
Total	178	(4,060)	1,550	(35,400)

[a]Summation of estimated municipal and rural groundwater withdrawals in 1960 and estimated additional withdrawals.

Source: Adapted from Bloyd (1974).

(230 Bgd), with the Ohio region ranking fourth in this respect (U.S. Geological Survey 1977).

Projections to the year 2000 indicate that development of energy resources (coal, oil, gas, oil shale, and tar sands) in the Ohio River basin could stress available water supplies. Monthly flow varies considerably in the Ohio River basin, and some upper mainstem and tributary areas already experience low-flow problems. Increased energy development in the Allegheny, Monongahela, Muskingum, Scioto, Miami, Hocking, and Kentucky river basins could exacerbate low-flow problems. Future energy demand has been seen as reducing the 7-day 10-year low-flow in the Ohio River mainstem by 15% in 2000.

Reservoir development (in addition to the importation of
power) has been identified as a possible future need (Water
Resources Council 1974; Peterson and Sonnichsen 1976).
Shepherd (1979) has examined with county-level resolution
projected water demands and availability in the Ohio River
basin and has assessed the water-use implications of energy
development in the basin. His analysis indicated possible
water availability problems in the Green, Monongahela, and
Conemaugh river basins by 1985, with significant water availability problems on the middle and lower Ohio River mainstem
projected for 2020.

Aquatic Ecology

     The Ohio Water Resource Region contains some of the same
habitats found in the Tennessee WRR. Mountain streams such as
those described in the section on the Tennessee region are
common in some parts of Pennsylvania, West Virginia, Tennessee,
Kentucky, and Ohio (Patterson 1970). Because these areas
are also heavily mined for coal, many of the same types of
mine drainage problems exist here. Although some of the more
remote streams have escaped degradation, much of the Appalachian portion of this region has been greatly impacted (Gerking
1966). Consequently, the natural biota has been largely
displaced by more tolerant forms. Thus, trout that were once
abundant locally have been virtually eliminated.
     The region contains several large rivers that, because of
their slow-flowing characteristics, more nearly resemble
turbid lakes, in many respects. The Ohio River, for example,
contains reproducing plankton populations in several areas
(Gerking 1966). A warmwater fishery of considerable sport
value occurs in many of these rivers, despite the problem of
progressive water quality degradation (Migdalski 1962). Fish
commonly caught include carp, catfish, drum, bass, and bluegill. Although commercial fishing (e.g., for walleye) once
occurred extensively on a few of the larger rivers, little is
done today (Gerking 1966). One of the greatest obstacles to
successful fishing in some areas is the presence of
bioaccumulated toxicants in fish tissue (e.g., mercury, DDT),
which has required limitations on the amount of fish a person
can safely consume (Hynes 1971).
     The upper one-third to one-half of this WRR was glaciated,
and most of the natural lakes occur in this area. Because of
the great numbers of people living nearby, these lakes receive
considerable fishing pressure and are constantly enriched with
anthropogenic nutrients (Geraghty et al. 1973). Some of the
more shallow lakes have become overgrown with macrophytes (as
a result of eutrophication) to the extent that frequent
anoxic conditions develop that cause many sensitive organisms

34

to disappear (Gerking 1966; Gerking 1950). Some progress has been made in alleviating this problem by increased sewage treatment capabilities, weed harvesting, and in situ nutrient inactivation (Hynes 1971).

The region contains several wetlands (particularly in the glaciated area). Marshes are most common, but some bogs occur in northern Indiana and Ohio (Geraghty et al. 1973).

UPPER MISSISSIPPI

Hydrology, Water Quality, and Water Use

The Upper Mississippi Water Resource Region, with an area of about 490,000 km^2 (190,000 sq miles) (U.S. Geological Survey 1977), includes the Mississippi from its headwaters in northcentral Minnesota downstream to the mouth of the Ohio River and intermediate tributary systems — the Minnesota, Wisconsin, Des Moines, and Illinois rivers. The region also includes the lowest reach of the Missouri River from its mouth near St. Louis, Missouri, upstream to the confluence of the Osage River. Natural lakes, found throughout the basin, are prominent features in central and northern Minnesota (such as Leech and Mille Lacs lakes), reflecting the glaciated history of the area. In addition to the large Red Rock Reservoir on the Des Moines River in Iowa, there are many smaller impoundments along other waterways.

Surface waters of the Upper Mississippi Water Resource Region are generally hard (120 to 180 mg/liter hardness as $CaCO_3$), with some areas of even greater hardness in central Illinois and in the western section of the region. Low concentrations of TDS (<120 mg/liter) are generally found only in western and northern Wisconsin. High levels of TDS (>350 mg/liter) are found in central and northeastern Illinois, central and northern Iowa, and southwestern Minnesota. Intermediate levels of TDS are found elsewhere in the region. Levels of total suspended solids are low (<270 mg/liter) in most of Wisconsin and Minnesota and along the mainstem of the Mississippi River as far downstream as St. Louis, Missouri. The highest levels (>1900 mg/liter) are reported from a belt of tributary areas along the mainstem, including parts of Wisconsin, Iowa, Illinois, and Missouri, and from northeastern Missouri and southeastern Iowa. Intermediate TSS levels (270 to 1900 mg/liter) are found in the remainder of Iowa, Missouri, and Illinois (Geraghty et al. 1973).

Water quality in the Upper Mississippi Water Resource Region has been adversely affected by a variety of point-source and nonpoint-source pollutants. Point-source pollutants, that is, industrial and municipal discharges, have caused elevated levels of heavy metals, such as mercury contamination in some Wisconsin streams and cadmium and chromium contamination in the Mississippi mainstem (Geraghty et al. 1973; U.S. Environmental Protection Agency 1977b). Papermill wastes have also degraded water quality in Wisconsin. Diffuse discharges, such as construction and urban and agricultural runoff, have increased the loads of total dissolved and suspended solids (U.S. Environmental Protection Agency 1977b). The effect of feedlots on levels of phosphorus and nitrogen in some Iowa streams has been described by Jones et al. (1976). They found significant correlations between animal units in feedlots and stream nutrient loads. The effect of land use on water quality is extensively described in U.S. Environmental Protection Agency (1976, 1977c). Pesticide residues have been detected in the mainstem of the Mississippi (Geraghty et al. 1973). The southwestern part of Illinois and, to a lesser extent, parts of Iowa and Missouri have experienced acid mine drainage, particularly after heavy rains and runoff (U.S. Environmental Protection Agency 1977b). The Upper Mississippi Water Resource Region has been categorized as having major thermal pollution problems, although not as severe as the Great Lakes, New England, or Mid-Atlantic regions (Geraghty et al. 1973). Brines from manufacturing processes and oil fields have caused local water quality degradation. The Rubicon River has had chloride levels exceeding 600 mg/liter (Water Resources Council 1974).

The Upper Mississippi Region has an estimated 1.7×10^{11} m^3 (45,000 $\times$ 10^9 gal) of potable groundwater in storage in outwash and alluvial aquifers alone; in addition, several times this amount is probably in storage in other types of aquifers (Bloyd 1975). The four major types of aquifers are Holocene (recent) alluvium, glacial outwash, buried valley deposits, and bedrock.

The alluvial aquifers are comprised of gravel, sand, and silt. These deposits are found in the Mississippi River valley and along the larger streams.

Outwash aquifers are somewhat more permeable than the alluvial aquifers, being composed of sand and gravel. These glacial deposits are most commonly found in Minnesota and Wisconsin. The alluvial and outwash aquifers range in thickness from 9 to 61 m (30 to 200 ft). They have the potential to yield more than 32 liters/sec (500 gpm) to wells.

Buried valley deposits are common in east-central Illinois, where at least 9600 km^2 (3700 sq miles) of area are known (Bloyd 1975).

The Upper Mississippi Region is underlain by bedrock of various ages. Three major bedrock aquifers exist: the Mount Simon–Hinckley, the Cambro–Ordovician, and the Silurian-Devonian. The Mount Simon–Hinckley, which is of Precambrian to Cambrian age, is a sandstone aquifer underlying southeastern Minnesota, western and southern Wisconsin, northern

36

Illinois, and eastern Iowa. The Cambro-Ordovician aquifer consists of a series of sandstones and dolomites. It underlies southeastern Minnesota, southern Wisconsin, northern Illinois, Iowa, northwestern Indiana, and eastern Missouri. The Silurian-Devonian aquifer is also predominantly limestone and dolomite. It is a source of water for northeastern Iowa, northern Illinois, southeastern Wisconsin, and northwestern Indiana. The bedrock aquifers have the potential for yielding as much as 32 liters/sec (500 gpm) to wells (Bloyd 1975).

Groundwater quality in the Upper Mississippi Region is generally good; the TDS concentrations are usually less than 1000 mg/liter. In the bedrock aquifers, however, saline water occurs at depth. The greatest TDS concentrations occur in southern and central Illinois and western Iowa, where deep sedimentary basins are present (Bloyd 1975).

Recharge to the various aquifers occurs through infiltration of precipitation, interaquifer flow, and from influent streams (Table VIII). The major discharge areas are the Mississippi River valley, Lake Michigan, pumpage, and the Illinois structural basin. The rate of groundwater movement through the region ranges from a meter to perhaps a hundred meters per year.

All base-year (1960) rural water was derived from groundwater reservoirs (Table IX). The annual domestic, commercial, and rural use of groundwater was approximately 3.4×10^6 m^3/day (900 Mgd), which is 4% of recharge [8.7×10^7 m^3/day (23,000 Mgd)]. Industrial use of groundwater was only 3% of recharge in 1965, or 2.4×10^6 m^3/day (629 Mgd) (Bloyd 1975).

In the Upper Mississippi Water Resource Region, average annual runoff is equivalent to 250×10^6 m^3/day (65 Bgd). Of the total off-channel water withdrawal of about 72×10^6 m^3/day (19 Bgd) in 1975, surface freshwater withdrawals accounted for 61×10^6 m^3/day (16 Bgd); groundwater contributed about 9.1×10^6 m^3/day (2.4 Bgd). About 4% of freshwater withdrawn was actually consumed. Public supplies and the self-supplied industrial sector were the primary users of groundwater. The greatest user of water in the region is the self-supplied industrial sector, including electricity-generating utilities (Table I). However, the greatest consumption of freshwater was accounted for by rural uses. Generation of hydroelectric power uses about 420×10^6 m^3/day (110 Bgd), with the Upper Mississippi region (along with the Arkansas-White-Red Water Resource Region) trailing eight other regions in this respect (U.S. Geological Survey 1977).

A projection of water-supply availability until the year 2020 has concluded that, on a regional basis, supplies will be adequate to accommodate expected demand. However, compared with most other eastern regions, precipitation and runoff are low and more variable. As a result, a large concentration of water demands could cause local low-flow problems, even on the

Table VIII. Summary of groundwater recharge computations

Subbasin	Subbasin area [km² (sq miles)]	Estimated groundwater recharge (m³/sec)	(Mgd)	Subbasin precipitation (%)
Mississippi headwaters	72,779 (28,100)	206	4,700	14
Chippewa-Black	33,929 (13,100)	96	2,200	12
Wisconsin	33,152 (12,800)	127	2,900	15
Rock	37,555 (14,500)	131	3,000	13
Illinois	76,664 (29,600)	127	2,900	6
Kaskaskia	18,130 (7,000)	31	700	6
Big Muddy	7,252 (2,800)	7	150	3
Meramee	18,130 (7,000)	55	1,250	9
Salt	11,655 (4,500)	7	150	2
Fox-Wyaconda-Fabius	7,770 (3,000)	4	100	2
Des Moines	38,332 (14,800)	31	700	3
Skunk	11,914 (4,600)	18	400	6
Iowa-Cedar	33,152 (12,800)	55	1,250	6
Turkey-Maquoketa-Upper Iowa-Wapsipinicon	22,792 (8,800)	61	1,400	10
Cannon-Zumbro-Root	14,763 (5,700)	35	800	10
Minnesota	43,771 (16,900)	24	550	3
Total		1,010	23,150	

Source: Adapted from Bloyd (1975).

Table IX. Base-year (1960) domestic, commercial, and rural groundwater use [m³/sec (Mgd)]

Subbasin	Domestic and commercial		Rural domestic and livestock		Total	
Mississippi headwaters	2.45	(56)	1.80	(41)	4.25	(97)
Chippewa-Black	0.48	(11)	0.96	(22)	1.45	(33)
Wisconsin	1.01	(23)	1.05	(24)	2.06	(47)
Rock	3.64	(83)	1.31	(30)	4.95	(113)
Illinois	6.75	(154)	3.29	(75)	10.03	(229)
Kaskaskia	0.39	(9)	0.48	(11)	0.88	(20)
Big Muddy	0.02	(0.4)	0.22	(5)	0.22	(5)
Meramec	0.39	(9)	0.48	(11)	0.88	(20)
Salt	0.09	(2)	0.48	(11)	0.57	(13)
Fox-Wyaconda-Fabius	0.03	(0.6)	0.22	(5)	0.26	(6)
Des Moines	1.01	(23)	2.10	(48)	3.11	(71)
Skunk	0.35	(8)	0.70	(16)	1.05	(24)
Iowa-Cedar	1.62	(37)	2.28	(52)	3.90	(89)
Turkey-Maquoketa-Upper Iowa-Wapsipinicon	0.66	(15)	1.58	(36)	2.23	(51)
Cannon-Zumbro-Root	0.66	(15)	0.74	(17)	1.40	(32)
Minnesota	0.70	(16)	1.53	(35)	2.23	(51)
Totals	20.24	(462)	19.23	(439)	39.47	(901)

Source: Adapted from Bloyd (1975).

mainstem of the Mississippi, until downstream tributary replenishment was sufficient (Water Resources Council 1974; Dobson and Shepherd 1979).

Aquatic Ecology

The most prominent water resource features of this WRR are the Mississippi River, several smaller rivers, and several lakes of glacial origin, particularly in the northern half of the region (Strahler and Strahler 1976). The trophic status of several lakes in this WRR is discussed in U.S. Environmental Protection Agency (1977d). This region contains the Mississippi River from its source in central Minnesota to its confluence with the Ohio River at the southern border of Illinois (Fig. I). Along this stretch, the river changes from a swift, relatively shallow stream to a large, slow-moving and turbid system. The biota in the upper reaches is markedly different from that downstream. In central Minnesota, the Mississippi River contains an abundant and diverse bottom fauna, largely because of the shallowness of the water and the abundance of dissolved oxygen. Downstream stretches have much less diverse communities, although some species (e.g., chironomids) are locally abundant (Eddy 1966; Carlander, Campbell, and Irwin 1966). The velocity of the river below central Minnesota has been further slowed by the construction of several impoundments. These have created pooled areas that resemble medium-hard-water lakes in the vicinity (Hynes 1972). Plankton populations, for example, are seasonally extensive in these areas, with diatoms and blue-green algae generally dominant (Eddy 1966). Fish in the backwater areas surrounding these pools are likewise more lacustrine than riverine, and centrarchids are common. In contrast, the more free-flowing downstream reaches include a typical large-river fauna of paddlefish, sturgeon, and species of *Carpiodes*, and the reproducing plankton population is relatively scant (Eddy 1966).
The Mississippi River receives large quantities of wastes downstream from the Minneapolis-St. Paul area (Geraghty et al. 1973). In terms of biotic effects, the greatest problem has been the addition of municipal wastes that have caused severe deoxygenation in some areas (Eddy 1966). The effect of the lowered oxygen levels has been to select for a more resistant biota; the most conspicuous change has been the greater proportion of rough fish that inhabit the river (Hynes 1972). Other rivers in the area have experienced similar problems, although in several the high-BOD source is papermill wastes. Agricultural runoff has likewise contributed greatly to the eutrophication of streams, and large algal blooms in surface waters have been common as a result of this input (Kilkus et al. 1975). Other rivers have received considerable inputs of

heavy metals that have bioaccumulated resulting in the limitation on human consumption of fish flesh from them (Geraghty et al. 1973; Eddy 1966). Mine drainage in parts of Illinois has greatly altered the biota of several streams there. In several areas, fish have been completely eliminated and only a highly resistant fauna and flora exist (Hynes 1971).

The natural lakes in this WRR occur mainly in the highly glaciated areas of central-northern Minnesota and Wisconsin (Patterson 1970). The lakes tend to be soft and unproductive in the eastern half of this area and hard and eutrophic to the west. The soft water lakes are similar to those described for northeast Minnesota in the section on the Souris-Red-Rainy region and the hard water lakes are essentially the same as those described for the southeastern portion of the Souris-Red-Rainy WRR (succeeding section). Most natural lakes that occur in the southern portion of the Upper Mississippi WRR are highly productive, and many contain large warmwater sport fisheries (Carlander, Campbell, and Irwin 1966; Gunning 1966). Anthropogenic eutrophication has degraded several of these lakes; mine drainage has greatly affected others, although the hard water of these lakes makes them somewhat resistant to the effects of moderate mine effluent inputs (Hynes 1971; National Academy of Sciences and National Academy of Engineering 1972; Carlander, Campbell, and Irwin 1966; Gunning 1966).

LOWER MISSISSIPPI

Hydrology, Water Quality, and Water Use

The Lower Mississippi Water Resource Region, with an area of about 250,000 km^2 (96,000 sq miles) (U.S. Geological Survey 1977), includes the lower mainstem of the Mississippi River, the largest river in North America, from southern Missouri to the delta in Louisiana. The region also includes the lower reaches of major tributary systems (Arkansas, Red, White, and Yazoo rivers) as well as smaller tributaries. Natural lakes are not important surface features of the region, except in coastal Louisiana. A noteworthy exception is Reelfoot Lake in northwestern Tennessee, which was formed by the New Madrid earthquake in the nineteenth century. Reservoirs are abundant in the region, however, including those in Missouri (Wappapello Reservoir), Arkansas-Louisiana (Bayou Bodcau Reservoir), and Mississippi (Arkabutla Reservoir). The Louisiana coast has an abundance of bayous, bays, and coastal lakes.

Surface waters in the Lower Mississippi Water Resource Region are generally of moderate hardness (60 to 120 mg/liter hardness as $CaCO_3$) in the Mississippi River mainstem, lower reaches of tributary areas, and the delta area. Exceptions are found (120 to 180 mg $CaCO_3$ per liter) below Baton Rouge,

Louisiana, and above the mouth of the Arkansas River. Otherwise, portions of the region more than about 80 km (50 miles) from the Mississippi River are typically soft (<60 mg/liter). Total dissolved solids along the Mississippi River mainstem, Red River, and in coastal Louisiana are generally in the 120 to 350 mg/liter range. Higher concentrations (>350 mg/liter) of TDS are found along the Ouachita and Arkansas rivers. Otherwise, TDS levels are typically low (<120 mg/liter). Loads of TSS are in the 270 to 1900 mg/liter range along the Mississippi River mainstem and in the upper Ouachita River basin. The highest levels (1900 mg/liter and above) are found in a belt of uplands running from southwestern to northern Mississippi. Otherwise, concentrations are low (<270 mg/liter) (Geraghty et al. 1973).

Nonpoint-source water pollution is a serious problem in the Lower Mississippi region, a problem not easily controlled. The importance of agricultural operations in the region (reflected in water use) largely determines the nature of water quality problems (U.S. Environmental Protection Agency 1977b). Agricultural runoff, especially runoff of bacteria, pesticides, and sediments, for example, may result in the failure of streams in Arkansas to meet water quality standards; pesticide residues have been found in the Mississippi mainstem and in lower tributary reaches (Geraghty et al. 1973; U.S. Environmental Protection Agency 1977b). Point sources, such as industrial and municipal waste discharges, may be controlled with waste treatment facilities; this has been shown, for example, in the control of municipal waste discharges by the state of Mississippi into the Mississippi River (U.S. Environmental Protection Agency 1977b). Major streams have been and are continuing to be contaminated with toxic elements such as arsenic, cadmium, and mercury (Geraghty et al. 1973). Surface waters in Arkansas have been adversely affected by acid drainage (from bauxite mines) and oil field brine point-source and diffuse releases (U.S. Environmental Protection Agency 1977b).

The annual amount of groundwater containing less than 3000 mg/liter TDS available from storage is 24×10^9 m^3 (19.5×10^6 acre-ft); total groundwater storage is about 9.8×10^{12} m^3 (8×10^9 acre-ft) (Terry et al. 1979). The aquifers consist of unconsolidated to semiconsolidated deltaic and fluvial sediments. The primary aquifers consist of Tertiary and Quaternary sand and gravel, although deposits of other ages are also used.

The major Cretaceous aquifer is the McNairy sand and its equivalents, which underlies an area of about 52,000 km^2 (20,000 sq miles), in the northern portion of the region. Other Cretaceous aquifers are the Coffee Sand and the Gordo Formation (Terry et al. 1979).

Important aquifers of Eocene age include the lower Wilcox Group, and the following units of the Claiborne Group: Carrizo Sand, Sparta Sand, and Cockfield Formation. The lower Wilcox aquifer is used in northeast Arkansas and northwest Mississippi. The Carrizo Sand is a source of water for central Arkansas and western Mississippi. In the northern 75% of the region, the Sparta Sand is quite productive. The Cockfield aquifer is used in portions of Louisiana, Arkansas, and Mississippi. The Memphis aquifer consists of the previously mentioned Claiborne formations, in a sequence almost one hundred meters thick (Terry et al. 1979).

The most extensive source of groundwater in the region is the Mississippi River valley alluvial aquifer. The major producing zones are gravel and sand deposits of Pleistocene age.

Well yields in the valley alluvial aquifer commonly exceed 100 liters/sec. Aquifers capable of producing at least 32 liters/sec (500 gpm) occur under 90% of the region. The groundwater quality is good; only a small portion of the region is underlain by aquifers with a TDS concentration of more than 1000 mg/liter. Saline water occurs at depth and with proximity to the mouth of the Mississippi River (Walton 1970). Precipitation on outcrop areas provides most aquifer recharge; discharge is to pumpage and streamflow (Terry et al. 1979).

In the Lower Mississippi Water Resource Region, average annual runoff is equivalent to 300×10^6 m^3/day (79 Bgd) (U.S. Geological Survey 1977). An additional 1330×10^6 m^3/day (352 Bgd) flows into the region from upstream areas (Water Resources Council 1974). Of the total off-channel water withdrawal of about 61×10^6 m^3/day (16 Bgd) in 1975, surface freshwater withdrawals accounted for about 42×10^6 m^3/day (11 Bgd); fresh groundwater accounted for most of the remainder. About 34% of freshwater withdrawn was actually consumed. The greatest user of water in the region is the self-supplied industrial sector, including electricity-generating utilities (Table I). However, the greatest freshwater consumption was accounted for by irrigation. Generation of hydroelectric power uses about 16×10^6 m^3/day (4.1 Bgd), making the region the fourth smallest user in the conterminous United States in this respect (U.S. Geological Survey 1977).

A projection of water-supply availabilities to the year 2000 indicates that the region will probably have adequate supplies for expected energy development (electrical generation and other technologies), even if the region must supply power to the urban centers of Arkansas and southeastern Texas. However, maintaining sufficient flows in the Mississippi River for saltwater repulsion at New Orleans, Louisiana, for navigation, and for environmental considerations may be a problem if large amounts of water are consumed at upstream reaches (Water Resources Council 1974; Dobson and Shepherd 1979).

Large natural lakes are not a prominent feature of this WRR, primarily because the area has not been glaciated (Strahler and Strahler 1976). Several impoundments occur in the area, however, and oxbow lakes, coastal lakes, and small ponds are common. In general, these waters are moderately hard and highly productive (Moore 1966). Many support active warmwater fisheries [the region has 0.62×10^6 ha (1.52×10^6 acres) of fishable freshwater], although the shallower ponds and lakes often suffer from deoxygenation with resultant frequent fish kills (Geraghty et al. 1973). Cultural eutrophication of these waters is a serious problem affecting the biota; the major nutrient additions come from fertilizer runoff and municipal wastes (Moore 1966). A biotic problem particularly acute in the extreme southern United States and greatly affecting this region is the introduction and success of exotic species, many of which are of tropical origin. The water hyacinth, for example, has virtually overgrown many of the smaller benthic environments, causing considerable habitat alteration. Dense mats of this species cause subsurface deoxygenation and dense shading, resulting in the degradation of the natural communities (Moore 1966).

Marsh and swamp lands are locally abundant in the lower Mississippi basin, particularly near the coastal mouths of rivers (Geraghty et al. 1973). Many of these have alternate freshwater—brackish-water intrusions, and the biota of the most seaward ones is very euryhaline (Reid 1961). Despite these stressful conditions, these habitats are highly productive; many serve as important nursery and spawning grounds for coastal marine organisms (Macan 1974). Thus, their ecological integrity is highly important to the region. Some of the inland wetlands are dystrophic and resemble northern bogs in their chemical makeup and low productivity (Moore 1966).

Although several large rivers exist in this WRR, the Mississippi is the most conspicuous one by far. In many respects its biota is much the same here as it is in the lower reaches of the Upper Mississippi WRR (see section immediately preceding) (Moore 1966). Although its volume increases markedly along its length, velocities are approximately the same as those encountered in southern Illinois because the elevational gradient from the Twin Cities to the Gulf of Mexico is slight (Strahler and Strahler 1976). Deoxygenation problems become greater along its course, however, as a result of additional inputs of sewage, fertilizer, etc. Similarly, heavy metal and pesticide accumulation in organisms generally increases downstream (Moore 1966). The higher silt load (largely from erosion) carried by the river in its lower reaches inhibits plankton production in some of the impounded areas, but a greater proportion of productive backwater areas makes up for this loss. Fish are locally abundant, especially near backwater areas, and sport fishing is common (Moore 1966).

ARKANSAS—WHITE—RED

Hydrology, Water Quality, and Water Use

The Arkansas—White—Red Water Resource Region, with an area of about 685,000 km^2 (265,000 sq miles) (U.S. Geological Survey 1977), includes all but the lowest reaches of the Arkansas, White, and Red river systems and their tributary systems (Cimmaron, Canadian, Washita, and Neosho rivers). Although the region does not have an abundance of natural lakes, most of the major streams have been impounded. Prominent reservoirs exist on the Red River (Lake Texoma), Arkansas River (Kaw and Keystone reservoirs), and White River (Beaver, Table Rock, Bull Shoals, and Norfolk reservoirs), in addition to those on tributaries (such as Eufala on the Canadian River and Lake of the Cherokees on the Neosho River).

Surface waters of the Arkansas—White—Red range from soft to moderately hard (<60 to 120 mg/liter hardness as $CaCO_3$) in the eastern parts of the basin (southeastern Oklahoma, southern Missouri, and Arkansas) to moderately hard and to hard (60 to 240 mg/liter) in the central and upper reaches of the Arkansas and Red river systems. The hardest surface waters are found in northeastern Oklahoma and southeastern Kansas. Total dissolved solids concentrations are typically high (>350 mg/liter) throughout the western parts of the region and in the lower mainstems of the Arkansas and Red rivers; portions of the upper Red River, Arkansas River, and Canadian River drainages have salinities exceeding 1000 mg/ liter. Areas in southeastern Oklahoma and in Arkansas have low levels of TDS (<120 mg/liter), while intermediate levels of TDS are found in the transition zone (eastern Oklahoma, southeastern Kansas, southern Missouri, northwestern Arkansas). Distribution of TSS concentrations exhibits a similar pattern: highest values (>1900 mg/liter) in the western parts of the region, low to intermediate values (<1900 mg/liter) in the eastern parts (Geraghty et al. 1973).

Point sources (such as along the Arkansas River mainstem in Kansas) have been responsible for part of the surface-water quality problems in the region. However, the nonpoint sources are more important on a regional basis and more difficult to control. These include sources such as urban—industrial runoff, agricultural runoff, and natural mineralization. Salt and gypsum formations in Oklahoma contribute to the minerali- zation of surface waters; in Kansas, stream water use is limited by natural mineralization. The addition of dissolved solids and nutrients contributes to the overall degradation of water quality from headwaters to lower reaches (U.S. Environmental Protection Agency 1977b). Pesticide residues and elevated trace element (such as lead, arsenic, and cadmium) concentrations have been reported in surface waters (Geraghty et al. 1973).

Aquifers in the Arkansas–White–Red region contain an estimated 2.46×10^{12} m^3 (2×10^9 acre-ft) of freshwater in storage (Bedinger and Sniegocki 1976). These aquifers can be classified into four general types: alluvial, carbonate and gypsum, sand and sandstone, and undifferentiated consolidated rocks (Table X).

Alluvial aquifers include stream valley, terrace, and intermontane valley deposits, consisting primarily of sand and gravel. The total area covered by stream valley alluvium alone is more than 72,520 km^2 (28,000 sq miles). Alluvial thicknesses range from 15 to 1500 m (50 to 5000 ft). Well yields vary between 3 and 3000 liters/sec (50 to 5000 gpm).

The sand and sandstone aquifers consist of unconsolidated sand along the Coastal Plain and consolidated sandstone underlying the High Plains and Central Lowland areas. Aquifer thickness is from 30 to 150 m (100 to 500 ft). Well yields typically range from 0.6 to 60 liters/sec (10 to 1000 gpm) (Bedinger and Sniegocki 1976).

The carbonate rock aquifers are comprised of dense limestones and dolomites of Paleozoic age. Outcrop areas include the Ozark Plateaus of Missouri and Arkansas. Gypsum aquifers occur in southwest Oklahoma and northern Texas. Permeability within these aquifers is secondary because of dissolution of the bedrock along joints and bedding planes. Thicknesses range from 15 to 460 m (50 to 1500 ft). Well yields are from 3 to 60 liters/sec (50 to 1000 gpm). Undifferentiated consolidated rock aquifers occur locally throughout the region. Thicknesses vary between 30 to 1500 m (100 to 5000 ft). Production is low; well yields typically do not exceed 3 liters/sec (50 gpm) (Bedinger and Sniegocki 1976).

Groundwater quality varies from fresh to saline within the region, both geographically and with depth. In general, TDS concentrations are lowest in the eastern part of the region. Saline groundwater (TDS concentration exceeding 1000 mg/liter) occurs at depths of less than 150 m (500 ft) over much of the region, as well as in areas underlain by shale and gypsum and where pollution by oil-field brine has occurred.

Recharge to the groundwater reservoirs varies across the region. In the semiarid western portion of the region, recharge may be as little as 0.13 to 1.3 cm (0.05 to 0.5 in.)/year. Recharge to aquifers in the humid eastern portion of the region may reach 51 cm (20 in.)/year (Bedinger and Sniegocki 1976). Discharge takes place through evapotranspiration and contribution to stream flow.

Table X. Principal aquifers in the Arkansas-White-Red region

Aquifer type	Nature of rock	Thickness (ft)	Areal extent	Depth to water (ft)	Hydraulic conductivity (ft/day)	Well yields (gpm)	Development and use	Groundwater in storage (acre ft $\times 10^8$)
Stream valley alluvium	Sand and gravel	50-200	Along large streams in flood plains. Extensive in Coastal Plain of Arkansas and Louisiana	0-30	100-1500	300-5000	Extensive; principal source of groundwater; frequently overdeveloped. Not used in some areas	2.8
Terrace alluvium	Sand and gravel	50-600	Plains of Texas, New Mexico, Colorado, Kansas, and Oklahoma	50-300	10-700	50-1000	Extensive subject to overdevelopment and water mining, particularly in High Plains of Texas	4.1
Alluvium of intermontane valleys and buried alluvial valleys	Sand and gravel	100-5000	Arkansas River basin in Colorado	0-50	10-700	50-1000	Extensive subject to overdevelopment and water mining, particularly in High Plains of Texas	0.2
Carbonate and gypsum	Limestone and dolomite and gypsum beds. Generally a dense rock, but subject to solution along fracture and bedding planes	50-1500	Limestone and dolomite in southern Missouri, northern Arkansas, southeastern Kansas, and Oklahoma. Gypsum in Oklahoma and Texas	30-450	50-1500	50-1000	Moderately to heavily developed; overlooked as a source of water in some areas. More subject to pollution than other aquifers because of cavernous nature	3.2
Sand and sandstone	Sand grains ranging from very fine to coarse. Generally cemented with siliceous material or carbonate. Unconsolidated in the Coastal Plain	100-500	Sandstone principally in Kansas, New Mexico, and Oklahoma Sand in Coastal Plain of Arkansas, Texas, and Louisiana	20-300	a	10-1000	Extensive; subject to overdevelopment and water mining. Loss of artesian head in many areas ranging from 2 to 300 ft	7.9

Table X (continued)

Aquifer type	Nature of rock	Thickness (ft)	Areal extent	Depth to water (ft)	Hydraulic conductivity (ft/day)	Well yields (gpm)	Development and use	Groundwater in storage (acre ft × 10^8)
Undifferentiated sandstone, carbonate, shale, or basalt	Consolidated rocks, including sandstone, interbedded shale, carbonate, and crystalline igneous rocks	100–5000	Sandstone, carbonate, and shale locally throughout region; basalt in parts of New Mexico, Colorado, and northwestern Oklahoma	1200	b	5–50	Mainly domestic use, not heavy, concentrated use, because of low permeability and low well yields. Difficult to predict well yields	2.2

[a]Generally less than 100 ft/day.
[b]Generally less than 10 ft/day.

Source: Bedinger and Sniegocki (1976).

Groundwater use in the Arkansas–White–Red region averaged 3.3×10^7 m^3/day (8.8 Bgd) (59% of total water use) in 1975. The primary use was for crop irrigation in portions of Texas, Oklahoma, Kansas, and Colorado. Other uses include providing rural and municipal supplies (Bedinger and Sniegocki 1976).

In the Arkansas–White–Red Water Resource Region, average annual runoff is equivalent to 280×10^6 m^3/day (73 Bgd). Of the total off-channel water withdrawal of about 57×10^6 m^3/day (15 Bgd) in 1975, surface freshwater withdrawals only accounted for about 24×10^6 m^3/day (6.2 Bgd); fresh groundwater contributed almost all of the remainder. Surface-water withdrawals are dominant in the eastern part of the region; groundwater withdrawals are dominant in the western and central parts. About 60% of freshwater withdrawn was consumed. Irrigation is clearly the dominant user and consumer of all waters in the region (Table I) (U.S. Geological Survey 1977). Imports from the Upper Colorado region are used to augment surface waters in the fully appropriated upper Arkansas River basin (Colorado, New Mexico) (Water Resources Council 1974). Generation of hydroelectric power uses about 420×10^6 m^3/day (110 Bgd), causing the Arkansas–White–Red Water Resource Region (along with the Upper Mississippi region) to trail eight other regions in this respect (U.S. Geological Survey 1977).

A projection to and beyond the year 1985 has indicated that water supplies in the Arkansas–White–Red region will be sufficient to meet expected energy needs (Water Resources Council 1974), although a more recent analysis (Dobson and Shepherd 1979), indicates potential problems in both ground and surface water availability. Areas of growing demand (the westernmost part of the region) do not have sufficient water supplies, but demands may be met by purchase of existing water outlets and by water development projects. Limitations on water use also exist because of natural and man-made pollutants. Local effects on groundwater levels and artesian pressures may result from increased pumping (Water Resources Council 1974).

Aquatic Ecology

Although few natural lakes exist in this region, it contains a diversity of aquatic habitats. As in the Texas Gulf WRR section, aridity, hardness, total dissolved solids, and turbidity generally increase westwardly.

The eastern portion of the region contains the Ozark highlands where numerous small streams arise from the uplands as runoff or artesian spring discharge (Carlander, Campbell, and Irwin 1966). These water bodies are typically clear, cool, and relatively swift and contain biota similar to that found in more northern trout streams (Hynes 1972).

49

The streams to the west that drain the prairie areas are markedly different. These are characterized by high turbidities, highly fluctuating flows, sandy bottoms, and a paucity of biota (Carlander, Campbell, and Irwin 1966). Several flow only within the sand streambed during the low water months, thus causing fish and other nonburrowing fauna to be stranded in the occasional pools that persist in riverbed depressions. Selective pressure for those organisms able to withstand periodic crowding and low dissolved oxygen levels has resulted (Carlander, Campbell, and Irwin 1966). Even so, numerous fish die each year in these pools, and virtually no sport fishery exists in rivers subject to extreme seasonal drawdown. Because of the unstable substrate, few benthic organisms exist. The high turbidities limit the development of primary producers, although diatoms become abundant in the wet sand during drawdown (Carlander, Campbell, and Irwin 1966).

Most of the larger rivers of the region originate in the mountains to the west and have somewhat more stable flow regimes (Strahler and Strahler 1976). Many of these are now heavily impounded and regulated (Geraghty et al. 1973). Nevertheless, substantial reservoir level variations are common. These variations have limited the ability of macrophytes to establish themselves in such lakes and have hampered the spawning success of shallow-water nest builders, such as centrarchids (Carlander, Campbell, and Irwin 1966). The new reservoirs nonetheless tend to be highly productive and provide a good warmwater fishery for the first few years of their existence (Macan 1974; Patriarche and Campbell 1958). After five years or so, however, several species tend to become overabundant, growth rates slow, and fishing success drops off unless stringent management is observed.

Although a few natural lakes exist in the region, numerous farm ponds have been constructed, particularly in Oklahoma. These ponds tend to be very eutrophic and often harbor large algal standing crops, including toxic bluegreens. Bass, bluegill, and bullheads are often abundant in these ponds, although winter deoxygenation with resultant fish kills is common (Carlander, Campbell, and Irwin 1966).

Silt pollution is perhaps the greatest problem affecting the aquatic biota in this region. Although silt additions to waterways occur as a natural phenomenon in waters that drain the fine prairie soils, agriculture and construction have greatly exacerbated the condition (Geraghty et al. 1973; Carlander, Campbell, and Irwin 1966). The main effects are decreased photosynthesis and production by the primary producers (with effects on the entire food chain), gill clogging of fishes, smothering of benthic organisms, and decreased feeding and reproductive ability for several of the fish species (Hynes 1971; Hynes 1972). Fertilizer runoff has contributed to eutrophication problems, particularly in the reservoirs and small ponds, and pesticides have been excessively bioaccumulated in some riverine fish (Carlander, Campbell, and Irwin 1966). Heavy use of the river waters for irrigation in the western portion of this region has caused increased total dissolved solids concentrations, which in some cases have changed the natural biota to a more euryhaline assemblage (see Texas Gulf WRR section) (Minckley 1973).

TEXAS-GULF

Hydrology, Water Quality, and Water Use

The Texas Gulf Water Resource Region, with an area of about 450,000 km^2 (175,000 sq miles) (U.S. Geological Survey 1977), includes river systems draining most of Texas and portions of southeastern New Mexico and western Louisiana. Drainage is towards the Gulf of Mexico, with the coastal area including bays at the mouths of the streams and behind barrier beaches. Prominent river systems include the Sabine, Trinity, Brazos, Colorado, and Nueces rivers. Major reservoirs are present in central and eastern Texas (Toledo Bend, Sam Rayburn, Whitney, and Belton reservoirs and Lakes Travis and Buchanan).

Hardness of surface waters generally increases in a southwestward direction from soft (<60 mg/liter hardness as $CaCO_3$) in the eastern parts of the region (Sabine and Neches river basins) to moderate (60 to 120 mg/liter) in the Trinty River basin to hard (120 to 180 mg/liter) in the Brazos, Colorado, and lower Nueces river basins to very hard (180 to 240 mg/liter) in the upper Nueces River basin of southwestern Texas. Levels of TDS in surface waters are highest (>350 mg/liter) along the coast and in the upper river basins of northwestern Texas. Moderate concentrations of TDS (120 to 350 mg/liter) are found in parts of eastern and central Texas, and lowest levels (<120 mg/liter) are found in the lower Neches River basin area in southeastern Texas (Geraghty et al. 1973). In general, water quality decreases with increasing aridity. Groundwater contributions high in salt and gypsum affect the upper Brazos River basin (Water Resources Council 1974). Levels of TSS generally range from low (<270 mg/liter) in the lower Neches mainstem to moderate (270 to 1900 mg/liter) in eastern, coastal, and northwestern Texas to high (>1900 mg/liter) in central Texas and the upper Nueces River basin (Geraghty et al. 1973).

Heavy metals (including mercury) and pesticides have been identified as pollutants present in Texas waters (Geraghty et al. 1973). Brine disposal has also created severe local salinity problems, although control measures have limited adverse effects (U.S. Environmental Protection Agency 1977b).

Saltwater intrusion in the coastal groundwater of east Texas and west Louisiana has resulted from groundwater pumping (Water Resources Council 1974).

Significant groundwater reservoirs underlie more than 80% of the land area in the Texas-Gulf region. The twelve regionally important aquifers containing water with TDS concentrations less than 3000 mg/liter are given below [data from Baker and Wall (1976)].

The Hickory aquifer underlies portions of the Edwards Plateau and Llano Uplift of central Texas, for a total area of 12,950 km^2 (5000 sq miles). The aquifer, which consists primarily of sand and sandstone, is more than 122 m (400 ft) thick, extending downward for nearly 1520 m (5000 ft). Larger capacity wells completed in the Hickory yield between 13 to 32 liters/sec (200 to 500 gpm).

The Ellenburger-San Saba aquifer, which surrounds the Llano Uplift, has a surface area of 10,360 km^2 (4000 sq miles). The aquifer consists of more than 305 m (1000 ft) of limestone and dolomite, extending to depths of 914 m (3000 ft). Yields of larger capacity wells reach 63 liters/sec (1000 gpm).

The Santa Rosa aquifer is located east of the High Plains and has a surface area of 2590 km^2 (1000 sq miles). The Santa Rosa, which produces at depths as great as 137 m (450 ft), is comprised of sand and gravel. The average yield of large capacity wells is 16 liters/sec (250 gpm).

The Trinity aquifer's primary area of use is the Dallas-Fort Worth region. The surface area of the aquifer is approximately 51,800 km^2 (20,000 sq miles). The maximum thickness of the interbedded sand, shale, and limestone units is 366 m (1200 ft); they extend to a maximum depth of 1060 m (3500 ft).

The Edwards-Trinity (Plateau) aquifer underlies portions of the Edwards Plateau for an area of 38,200 km^2 (15,000 sq miles). The major producing zones are sand, sandstone, and honeycombed limestone; the maximum thickness is 305 m (1000 ft). Well yields can exceed 190 liters/sec (3000 gpm).

The Edwards (Balcones Fault Zone) aquifer lies between the Edwards Plateau and the Gulf Coastal Plain, along the Balcones Escarpment, for an area of 6470 km^2 (2500 sq miles). A 152-m (500-ft) thick zone of fractured limestone and dolomite provides the entire water supply for metropolitan San Antonio, Texas. Some wells yield more than 1000 liters/sec (16,000 gpm).

The Woodbine aquifer underlies an area of 15,500 km^2 (6000 sq miles) near the inland extent of the Coastal Plain. The aquifer is comprised of a 183-m (600-ft) thick sequence of sand, sandstone, and shale, extending to a depth of 610 m (2000 ft). Maximum well yields are 44 liters/sec (700 gpm).

The Carrizo-Wilcox aquifer has a total area of 77,700 km^2 (30,000 sq miles), bounding the Coastal Plain in a band 48 to 129 km (30 to 80 miles) wide. The interbedded sand and clay

reaches a thickness of 914 m (3000 ft). Some large capacity
wells produce as much as 190 liters/sec (3000 gpm).

The Queen City aquifer has an effective surface area of
36,300 km^2 (14,000 sq miles), underlying portions of the
Coastal Plain. Fresh water is produced down to 610 m
(2000 ft) from a 152-m (500-ft) thick zone of interbedded sand
and clay. Most well yields are considered to be low; however,
some exceed 25 liters/sec (400 gpm).

The Sparta aquifer underlies the Queen City aquifer and
overlies the Carrizo-Wilcox aquifers for an area of 23,300 km^2
(9000 sq miles). The production zone is an interbedded
sequence of sand and clay up to 107 m (350 ft) thick, occurring to a maximum depth of 610 m (2000 ft). Yields generally
range between 32 to 63 liters/sec (500 to 1000 gpm).

The Gulf Coast aquifer lies beneath the Coastal Plain
inland for up to 193 km (120 miles). It is the region's most
extensive groundwater reservoir, underlying an area of
90,600 km^2 (35,000 sq miles); it consists of a thickness of
more than 1060 m (3500 ft) of sand, clay, and gravel. The
Gulf Coast aquifer is most extensively developed in the
Houston area, where well yields average 126 liters/sec
(2000 gpm).

The Ogallala aquifer underlies 49,200 km^2 (19,000 sq
miles) of the High Plains. Major production comes from a
152-m (500-ft) thick zone of sand and gravel. Depths to water
range from 15 to 91 m (50 to 300 ft). Well yields vary from
6.3 to 63 liters/sec (100 to 1000 gpm) (Baker and Wall 1976).

In the Texas-Gulf region, groundwater quality varies with
depth as well as geographical location. Generally, the
lowest concentrations of dissolved solids occur in areas of
greatest rainfall and least evaporation. Table XI shows the
range in TDS concentrations for the regional aquifers.

Potential recharge to groundwater reservoirs in the
region increases from northwest to southeast, ranging from a
fraction of an inch on the High Plains to several inches in
the Sabine River basin. The calculated steady-state yields
(Table XII) are equivalent to the maximum annual recharge for
each aquifer. The partly recoverable water in storage is in
the form of freshwater bounded by saline water, which may or
may not be economically practical to recover (Baker and Wall
1976).

In the Texas Gulf region, average annual runoff is
equivalent to 120 × 10^6 m^3/day (32 Bgd). Of the total off-
channel water withdrawal of about 83 × 10^6 m^3/day (22 Bgd) in
1975, surface freshwater withdrawals only accounted for
approximately 37 × 10^6 m^3/day (9.7 Bgd); fresh groundwater
[27 × 10^6 m^3/day (7.2 Bgd)] and saline surface water
[19 × 10^6 m^3/day (5.1 Bgd)] contributed the remainder. About
47% of freshwater withdrawn was consumed (U.S. Geological
Survey 1977).

Table XI. Typical range of dissolved solids in
water used from each aquifer

Aquifer	Typical range in dissolved solids (mg/liter)
Alluvium	500–2000
Ogallala	400–1200
Gulf Coast	300–1000
Sparta	200–800
Queen City	200–800
Carrizo-Wilcox	200–1500
Woodbine	500–1200
Edwards (Balcones Fault Zone)	300–1200
Edwards-Trinity (plateau)	400–1000
Trinity	500–1500
Santa Rosa	400–2500
Ellenburger-San Saba	400–2000
Hickory	300–700

Source: Baker, Jr. and Wall (1975).

The self-supplied industrial and electricity-generation utilities sector accounts for most of the regional water use, although irrigation clearly accounts for most of the fresh-water consumption (Table I). About 85% of the irrigation demand is satisfied with groundwater (U.S. Geological Survey 1977). Groundwater in the northwestern part of the region (upper Brazos and Colorado basins) has been heavily pumped for agricultural needs. As a result, lowering of the water table has occurred — discharge exceeding recharge (i.e., the ground-water has been "mined"). Groundwater is also used for domestic, public, industrial, and municipal supplies and for secondary oil recovery throughout the region (Water Resources Council 1974). Saline surface-water use in the region is for the self-supplied industrial sector; cooling of thermoelectric facilities is responsible for about 55% of this use. Generation of hydroelectric power uses about 68×10^6 m^3/day (18 Bgd), with the Texas Gulf region ranking ahead of only five other regions in the conterminous United States in this respect (U.S. Geological Survey 1977). In a projection of future water supplies for energy development, the Water

Table XII. Quantities of groundwater available for development

Aquifer	Steady-state yield [10^5 m^3 (10^3 acre-ft)]		Recoverable water in storage above depths of 122 m (400 ft) [10^5 m^3 (10^3 acre-ft)]		Partly recoverable water in storage below depth of 122 m (400 ft) [10^5 m^3 (10^3 acre-ft)]	
Alluvium	1.6	(130)	61.7	(5,000)	0	(0)
Ogallala	1.1	(90)	1,660	(135,000)	61.7	(5,000)
Gulf Coast	31	(2,500)	5,550	(450,000)	14,200	(1,150,000)
Sparta	1.6	(130)	247	(20,000)	801	(65,000)
Queen City	1.5	(120)	863	(70,000)	2,470	(200,000)
Carrizo-Wilcox	6.9	(560)	1,850	(150,000)	14,200	(1,150,000)
Woodbine	0.1	(10)	123	(10,000)	863	(70,000)
Edwards (Balcones Fault Zone)	5.1	(410)	24.7	(2,000)	160	(13,000)
Edwards-Trinity (plateau)	6.7	(540)	863	(70,000)	863	(70,000)
Trinity	0.8	(70)	1,230	(100,000)	5,550	(450,000)
Santa Rosa	0.4	(30)	98.6	(8,000)	0	(0)
Ellenburger-San Saba	0.2	(20)	98.6	(8,000)	148	(12,000)
Hickory	0.5	(40)	123	(10,000)	1,230	(100,000)
Total (rounded)	57	(4,650)	12,800	(1,038,000)	40,500	(3,285,000)

Source: Adapted from Baker and Wall (1976).

Resources Council assessed the ground and surface waters of
the region to be generally adequate to and beyond 1985;
current and planned water development projects were taken into
account (Water Resources Council 1974). A more recent
analysis (Dobson and Shepherd 1979), however, indicates
potential water-supply problems throughout the region.

Aquatic Ecology of the Southwestern United States

    The southwestern portion of the United States contains a
variety of limnological features, but few of them have been
studied extensively (Cole 1966). The eastern portion of the
area is characterized by relatively high rainfall and low
evaporation; in general, the aridity increases westward as do
water hardness and total dissolved solids (Strahler and
Strahler 1976). In eastern Texas, bog remnants occur that are
similar in several respects to their northern counterparts.
Likewise, where rainfall is sufficient and drainage is
impeded, mountainous areas within these regions occasionally
contain bogs (Cole 1966).
    Reservoirs are a conspicuous aquatic resource in the
Southwest. Most have highly variable physicochemical conditions, because of the variability of precipitation. In
general, they are warm, monomictic water bodies with relatively high Na^+, Cl^-, $SO_4{}^{2-}$, and Ca^{2+} levels (Harris and
Silbey 1940). The plankton is generally scant, although
dinoflagellate or blue-green algal blooms sometimes occur in
late summer. *Daphnia* is usually the most common cladoceran,
and warm-water fish (e.g., centrarchids, shad, carp) are
prevalent. Trout are frequently found in tailrace waters.
Macrophyte communities are poorly developed, largely because
of the fluctuating water levels, and benthic standing crops
are generally less than 100 organisms per square meter (Cole
1966).
    Natural lakes in these regions are relatively scarce
(Strahler and Strahler 1976). Solution basins (largely in
gypsum deposits) sometimes contain water and often harbor
large populations of a few species. Particularly common are
Hyalella azteca and marginal stands of *Potamogeton* spp. (Cole
1966). Several mountain lakes exist; these are generally
dimictic and relatively unproductive. However, many contain
sizable trout populations and diverse invertebrate assemblages
(Juday 1907). Volcanic lakes occurring on the Colorado
Plateau contain a diverse biota, including considerable macrophyte development (Cole 1966). Playas and other ephemeral
lakes are common in certain areas and contain a specialized
biota highly resistant to desiccation. Algae and zooplankton
are seasonally numerous in these lakes, as are some aquatic
insect larvae (Macan 1974; Reid 1961).

Springs occur throughout these regions. Most are very high in TDS and contain little plankton. Mats of filamentous green algae are common, however, and ostracods, snails, and fly larvae are prevalent. Many of these springs contain a highly endemic fauna; several of their fish, in particular, are on endangered species lists (Cole 1966).

Several large rivers occur in these water resource regions. Those in eastern Texas are the lowest in dissolved solids and are more typical of rivers in Louisiana than of those farther west (Cole 1966). Coastal bays, estuaries, and swamps in Texas are also much like those in the Lower Mississippi WRR. The rivers in the more arid areas of the Southwest contain high dissolved solids levels and high turbidities. Benthic organisms are largely restricted to peripheral areas where scouring is not excessive; most fish spawning and plankton production also occur in these areas. The sport fishery of these rivers is substantial and is dominated by warmwater species, except in tailrace areas and mountainous tributaries (Minckley 1973).

Extensive irrigation canals occur in the Southwest. Most ditches that return water from the fields have increased salinities from evaporation and soil salt solubilization (Strahler and Strahler 1976). Although these can support a diverse freshwater biota, including a productive warmwater fishery, some of the more saline ditches develop substantial populations of euryhaline algae and other brackish-water species (Reid 1961; Minckley 1973).

Anthropogenic disturbances have greatly altered the aquatic habitats of this region. Eutrophication of surface waters from fertilizer and feedlot runoff has been acute (Geraghty et al. 1973). Salinity increases caused by agricultural operations have modified biotic communities, and pesticide additions have resulted in bioaccumulation and direct toxicity in biota (Cole 1966). Probably the greatest impact on biota has resulted from the extensive damming of the rivers, causing altered flow characteristics that favor species preferring slower velocities. As a result, several endemic species inhabiting these rivers have experienced declining populations (Minckley 1973). Numerous exotic species have been introduced into waters in the Southwest, many of which have become well established at the expense of the indigenous biota (Minckley 1973).

RIO GRANDE

Hydrology, Water Quality, and Water Use

The Rio Grande Water Resource Region, with an area of about 350,000 km^2 (136,000 sq miles) (U.S. Geological Survey 1977), includes the basin of the Rio Grande from headwaters in

57

southern Colorado to its mouth on the Gulf at the U.S.-Mexican border. Prominent reservoirs along the Rio Grande include Elephant Butte and Falcon reservoirs and Devils Lake. The Pecos River, originating in New Mexico, is the most important tributary to the Rio Grande.

The hardness of surface waters in the Rio Grande region ranges from moderate (60 to 120 mg/liter hardness as $CaCO_3$) in the headwaters of southern Colorado to hard (120 mg/liter) in central New Mexico and to very hard (>180 mg/liter) in southern New Mexico and Texas. The hardest waters (>240 mg/liter) are found in the Pecos drainage from the headwaters in New Mexico to its confluence with the Rio Grande and in the Rio Grande mainstem below the mouth of the Pecos River. The lowest levels of TDS (<120 mg/liter) in the region are found in the Rio Grande headwaters in northern New Mexico and southern Colorado. Downstream from Albuquerque, New Mexico, however, high levels of TDS (>350 mg/liter) predominate. Salinities exceeding 1000 mg/liter may be found in the Rio Grande mainstem below Albuquerque, New Mexico, downstream to the mouth of the Pecos River and in the Pecos River mainstem over almost its entire length. Concentrations of TSS in headwaters and tributaries of the Rio Grande and Pecos rivers and in reaches of the Rio Grande mainstem along portions of the Mexican border generally exceed 1900 mg/liter. Moderate levels (270 to 1900 mg/liter) are found in the mainstem of the Rio Grande from southern New Mexico downstream to the Big Bend, in the lower Pecos River mainstem, and in the lowest reaches of the Rio Grande downstream from Laredo, Texas (Geraghty et al. 1973). Surface water quality problems in the region include reservoir eutrophication (such as in Elephant Butte Reservoir) from municipal discharges and pesticides and increased salinities from agricultural sources (U.S. Environmental Protection Agency 1977b).

Aquifers in the Rio Grande region contain approximately 7.1×10^{12} m^3 (5800 $\times$ 10^6 acre-ft) of fresh-to-slightly-saline water in storage. The groundwater reservoirs can be classified according to four major types: valley fill (primarily unconsolidated to semiconsolidated sand and gravel), volcanic rocks, consolidated sedimentary deposits (shale, sandstone, limestone, gypsum, and salt), and crystalline rocks. The following data are from West and Broadhurst (1975).

The most important of the groundwater reservoirs is the valley fill material, which is found in intermontane valleys in all but the southeastern portion of the region. Sediment thicknesses range from very thin at valley perimeters to 2700 m (9000 ft); locally, deep basins with accumulations of more than 9100 m (30,000 ft) of sediments are encountered. Well yields are good and can be as great as 151 liters/sec (2400 gpm).

The volcanic rocks form the caprock for the plateaus, generally lying above the regional water table. They are only significant as aquifers where they occur in association with valley fill material.

Consolidated sedimentary rocks are found in the east-central and southeastern portions of the region, where they form most of the hills and low mountains. These units function as aquifers where fractures and solution openings occur within the bedrock. Well yields are generally in excess of 20 liters/sec (300 gpm); however, yields of 60 to 220 liters/sec (1000 to 3500 gpm) are common.

Crystalline (intrusive igneous) rocks crop out in the north and north-central portions of the region in the mountainous areas. The rocks are very dense and evidently not extensively fractured because well yields are generally insignificant.

Groundwater quality varies from site to site. Water produced from valley fill aquifers has TDS concentrations between 52 and 13,800 mg/liter; however, in some closed basin areas where evaporite deposits exist, the TDS concentration of the groundwater can be as great as 100,000 mg/liter. The quality of groundwater derived from the consolidated sedimentary rock aquifers usually ranges from less than 1000 to more than 35,000 mg/liter.

Recharge to the groundwater reservoirs takes place through infiltration of irrigation water, precipitation, and snowmelt. Discharge is by pumpage, evapotranspiration, seepage to streams, and minor underflow to adjacent water resource regions.

The average groundwater withdrawal in the region was 9.1×10^6 m^3/day (2.4 Bgd) in 1975. Approximately 90% of this total was used for irrigation. An additional 5% provided public supplies; the entire public water demands of Albuquerque and El Paso are met by groundwater withdrawal (West and Broadhurst 1975).

In the Rio Grande Water Resource Region, average annual runoff is equivalent to 20×10^6 m^3/day (5.4 Bgd). Of the total off-channel water withdrawal, about 20×10^6 m^3/day (5.4 Bgd) in 1975, surface freshwater withdrawals accounted for about 10×10^6 m^3/day (3 Bgd). About 65% of the freshwater withdrawn was consumed. Irrigation accounts for most of the total regional water use and freshwater consumption (Table I). Generation of hydroelectric power uses about 4.6×10^6 m^3/day (1.2 Bgd); the Rio Grande region ranks second to last in this respect (U.S. Geological Survey 1977). Rio Grande waters are divided between the United States and Mexico by treaty and between Colorado, New Mexico, and Texas by compact. It is apparent from runoff and water-use data that the region does not have sufficient water resources to support present development, and the Water Resources Council has listed the Rio Grande region as one of the nation's most

59

critical energy-related water-supply problem areas. Diversion of water from the Colorado River basin to the Rio Grande region via the San Juan—Chama Project augments regional supplies. Although a projection of water supply availability to the year 1985 indicates that energy development may be satisified by existing utility water rights and by purchase of additional existing water rights, increased competition for water is expected (Water Resources Council 1974).

Aquatic Ecology

The discussion of biotic resources in the Texas Gulf WRR section is applicable to the Rio Grande Water Resource Region.

SOURIS—RED—RAINY

Hydrology, Water Quality, and Water Use

The Souris—Red—Rainy Water Region, with an area of about 150,000 km^2 (59,000 sq miles) (U.S. Geological Survey 1977), includes areas of northern Minnesota, northeastern South Dakota, and northern and eastern North Dakota, all of which are drained ultimately into the Hudson Bay via Lake Winnipeg in Manitoba. The Red River basin includes headwaters as far south as Lake Traverse on the South Dakota—Minnesota border. The mainstem forms much of the North Dakota—Minnesota border, with drainage northward toward Lake Winnipeg. The Souris River originates in Canada, flows southward, and drains a portion of northern North Dakota, and then flows northward again into Canada, joining the Assiniboine River, a tributary to the Red River. The Rainy River drains a lake-filled region along the Minnesota-Ontario border, connecting Rainy Lake with the downstream Lake-of-the-Woods; the Rainy River mainstem itself forms part of the international border. Drainage from the Lake-of-the-Woods is to Lake Winnipeg via the Winnipeg River. Important tributaries include River Des Lacs (to the Souris River), Sheyenne, Bois de Sioux, and Red Lake rivers (to the Red River), and Big Fork and Little Fork rivers (to the Rainy River). The Souris—Red—Rainy region includes numerous large natural lakes (such as Lake-of-the-Woods, Rainy Lake, Upper and Lower Red lakes, and Lake Traverse), in addition to many smaller ones. Major reservoirs are not common in the region, notable exceptions being Lake Darling on the Souris River and Lake Ashtabula on the Sheyenne River. Surface waters are very hard (>180 mg/liter hardness as CaCO$_3$) in the western half of the region (extreme western Minnesota, the Red River mainstem, and the Souris River basin), with levels exceeding 240 mg/liter in North Dakota.

The Rainy River basin has moderately hard (60 to 120 mg/liter) waters; surface waters in the area in northern Minnesota tributary to the lower Red River are very hard (120 to 180 mg/liter). Lowest TDS levels are found in the Rainy basin above Rainy Lake (<120 mg/liter); moderate TDS levels (120 to 350 mg/liter) are found in the lower Rainy basin and in some areas of western Minnesota tributary to the Red River. Otherwise, high levels of dissolved solids (>350 mg/liter) are found. Some saline surface waters (TDS >1000 mg/liter) are found in North Dakota's central drift plain in the Devil's Lake area. Levels of TSS in streams of the region are generally low (<270 mg/liter) in the Rainy basin, upper Red River basin, and in the portion of the Sheyenne River affected by the impoundment of Lake Ashtabula; moderate concentrations (270 to 1900 mg/liter) are found elsewhere (Geraghty et al. 1973).

Reported toxic pollutants in region surface waters include pesticides (from the Rainy and Red rivers) and cadmium (from the Red River) (Geraghty et al. 1973). The Red River has been identified by the EPA as an example of stream recovery following waste discharge control — a stretch downstream from Fargo-Moorhead, North Dakota, had been severely impacted by municipal and industrial discharges from those cities. Implementation of waste treatment resulted in a significant improvement in water quality (U.S. Environmental Protection Agency 1977b).

The Souris-Red-Rainy region has an estimated 6.2×10^{11} m^3 (5×10^8 acre-ft) of water with a TDS concentration of less than 3000 mg/liter available from aquifer storage (Reeder 1978). This groundwater is obtainable from consolidated rock and glacial deposit aquifers. The primary aquifers in the region are glacial drift, the Fort Union, Fox Hills-Hell Creek, Pierre, Dakota, and undifferentiated Precambrian and Paleozoic rocks (Table XIII). The data below are from Reeder (1970).

Precambrian crystalline rocks provide groundwater locally where they have been fractured extensively or deeply weathered. Well yields are usually no more than a few tenths of a liter per second.

The undifferentiated units comprising the Paleozoic aquifer consist of fine-grained sandstone and limestone with solution openings. Yields are generally only several liters per second.

The Dakota Sandstone aquifer underlies the central portion of the region. Individual sand beds reach a thickness of 30.5 m (100 ft). Yields are highly variable, but are greatest in the Souris River sub-basin.

The Pierre Shale aquifer supplies groundwater from fractured zones. Yields are small but increase proportionately with degree of fracturing.

Table XIII. Aquifers and well yields in the Souris-Red-Rainy region

Aquifer	Well yields [liters/sec (gpm)]		
	Common rate pumped	General range	Largest known
Drift	<6.31 (<100)	0.32-63.1 (5-1000)	>63.1 (>1000)
Fort Union	0.13-0.25 (2-4)	<0.06-3.15 (<1-50)	6.31 (100)
Fox Hills-Hell Creek	<0.32 (<5)	<0.06-1.89 (<1-30)	9.48 (150)
Pierre	<0.32 (<5)	<0.06-0.38 (<1-6)	6.31 (100)
Dakota	0.31-0.19 (2-3)	<0.06-22.1 (<1-350)	31.55 (500)
Paleozoic	<0.32 (<5)	<0.06-3.79 (<1-60)	44.16 (700)
Precambrian	<0.32 (<5)	<0.06-0.63 (<1-10)	

Source: Adapted from Reeder (1978).

The Fox Hills-Hell Creek aquifer underlies the western portion of the basin. The aquifer, which consists of sand, clay, lignite, and silt, produces water from the thicker (on the order of 5 to 10 m) sand units. Well yields usually do not exceed 2 liters/sec.

The Fort Union aquifer underlies the westernmost portion of the Souris River sub-basin. The aquifer is composed of silt, clay, sand, and lignite. The best producing zones are thick (tens of meters) sand beds towards the top of the formation (Tongue River Member). Yields are on the order of several liters per second.

Valley fill or outwash deposits account for the bulk of the glacial drift aquifers; they comprise the most important groundwater sources in the region. Where the drift is stratified sand and gravel and where thicknesses are maximal (about 76 m or 250 ft), yields may exceed 63 liters/sec (1000 gpm).

Groundwater quality is generally good (Table XIV); groundwater containing less than 3000 mg/liter TDS is available throughout the region although it varies within and between aquifers. In recharge areas the drift aquifer water has less than 1000 mg/liter TDS; in many places TDS concentrations are lower than 500 mg/liter (Reeder 1978).

In the Souris—Red—Rainy Water Resource Region, average annual runoff is equivalent to 24×10^6 m^3/day (6.2 Bgd). Of the total off-channel water withdrawal of about 1.4×10^6 m^3/day (360 Mgd) in 1975, surface freshwater withdrawals accounted for about 1.0×10^6 m^3/day (270 Mgd) and fresh groundwater for about 0.3×10^6 m^3/day (90 Mgd). About 25% of freshwater withdrawn was consumed (U.S. Geological Survey 1977). The primary user of water in the region is the self-supplied industrial and electricity-generation utilities sector; most of the sector's withdrawals are for thermo-electric power generation (Table I). The primary consumer of freshwater, however, is irrigation. The region does not have major hydroelectric power capacity, making it unique among the regions of the United States in this respect (U.S. Geological Survey 1977).

Streamflow in the Souris—Red—Rainy region is shared with Canada by agreement. That part of the region located in North Dakota, which has localized water shortage and thermal pollution problems, has been identified by the Water Resources Council as one of the most critical energy-related water-supply problem areas in the nation. Mining and use of western North Dakota lignite coal reserves would require the development of additional storage or groundwater sources (Water Resources Council 1974).

63

Table XIV. Description of aquifers and water quality in the Souris–Red–Rainy region

Aquifer	Character of rocks	Water quality
Drift and alluvium	Drift: till, outwash, and lacustrine deposits Alluvium: gravel, sand, silt, and clay	Extremely variable from place to place. Dissolved solids generally 250 to 1,000 mg/liter, but may be as high as 3,000 to 10,000 mg/liter in a few places
Upper part of Fort Union (includes Tongue River Member)	Sandstone, shale and lignite	Sodium bicarbonate type and locally high sulfate. Dissolved solids 200 to 7,000 mg/liter
Cannonball Member of Fort Union	Marine sandstone and sandy shale	Sodium chloride type. Dissolved solids less than 3,000 mg/liter, most water ranges from 1,000 to 2,500 mg/liter
Fox Hills–Hell Creek	Sandstone; sandstone and shale	Sodium bicarbonate type generally, sodium sulfate type locally. Dissolved solids 300 to 3,700 mg/liter, but most water contains 1,000 to 2,000 mg/liter
Pierre	Shale, brittle, fissile; upper part fractured	Extremely variable from place to place. Sodium chloride, sodium sulfate type, or combination. Dissolved solids 700 to 12,500 mg/liter but most is less than 3,000 mg/liter
Dakota	Sandstone and shale	Souris basin: sodium chloride type, dissolved solids 4,000 to 15,000 mg/liter. Red basin: sodium sulfate type, dissolved solids 2,000 to 8,000 mg/liter in North Dakota, less than 2,000 mg/liter in Minnesota

Table XIV (continued)

Aquifer	Character of rocks	Water quality
Paleozoic	Shale, sandstone, lime-stone, and dolomite	Highly saline. Dissolved solids 14,000 to 54,000 mg/liter in central part of Red basin, and as high as 330,000 mg/liter in western North Dakota
Precambrian	Crystalline rocks, upper part fractured and weathered	Sodium bicarbonate sulfate type, depending largely on type in the adjacent aquifer. Dissolved solids about 1,000 mg/liter or more in Red basin, less than 500 mg/liter in Rainy basin

Source: Adapted from Reeder (1978).

This WRR contains a wide variety of aquatic habitats because it emcompasses serveral distinctly different physio-graphic features. The eastern portion of the region, which contains Precambrian bedrock with an overlay of nutrient-poor glacial drift, receives a relatively large amount of precipitation. The surrounding natural vegetation is predominantly coniferous forest. In contrast, the western portion of the WRR, which contains substrates that are much higher in soluble minerals, receives much less precipitation, the surrounding vegetation is largely prairie, and the drainage system is developed better because the area has been unglaciated for a longer period of time (Patterson 1970). Consequently, soft water lakes, bogs, and trout streams typify the eastern area, and saline, high-sulfate lakes typify the west (Eddy 1966).

The bog lakes and marshes of the eastern portion of this WRR are very similar to those described in the section on the New England WRR. The soft water lakes of northeast Minnesota are likewise similar to those found in Maine. In general, these lakes are characterized by narrow littoral zones (because of steep, rocky shorelines), very low natural productivity, little macrophyte production, and little plankton production (Eddy 1966). The plankton is dominated by diatoms (*Tabellaria, Fragilaria,* and *Asterionella* predominantly) and copepods. Profundal bottom areas often display large densities of *Pontoporeia affinis*, whereas the littoral benthic fauna is scarce. The deeper lakes contain cold-water fish species such as lake trout, ciscoes, and whitefish. Northern pike, yellow perch, burbot, and walleye are also found. These species are dominant in the shallower lakes that cannot support a cold-water fish because of to dissolved oxygen (DO) values during the summer (Eddy 1966). The trophic status of selected lakes in this WRR has been described in U.S. Environmental Protection Agency (1977d).

The southeastern portion of the WRR contains hard water, eutrophic lakes, and several marshy areas. Many of the lakes display summer anoxia in the hypolimnia; thus, a resistant profundal benthic fauna of chironomids and oligochaetes is common (Dineen 1953). The littoral areas of these lakes are highly developed; the standing crop of summer fauna is roughly ten times that of the soft water lakes of northeast Minnesota (Eddy 1966). Plankton are abundant, with blue-green algae dominant during the summer months. Most of the lakes support large fish populations (Dineen 1953). The larger bodies of water are dominated by walleye; the medium-sized by bass, crappie, and sunfish; and the smaller lakes by bullhead and carp. The lakes produce, on the average, 90 to 160 kg of fish per hectare (summer standing crop), and some of the shallower ones have been estimated to produce greater than 400 kg of fish per hectare. These estimates compare to an average of

35 kg of fish per hectare for fish standing crops in nine oligotrophic lakes in the extreme eastern portion of the WRR (Eddy 1966).

The prairie lakes within this WRR are high in sulfates, carbonates, chlorides, sodium, and calcium. Their salinity is highly variable. Some are more saline than ocean water, although the ionic proportions are usually quite different (Wilson 1958). Commonly, these environments are highly buffered and display pH values of 8.4–9.0 (Reid 1961). Brackish-water emergent plants are common, with *Ruppia occidentalis* and *Nais marina* often predominating (Eddy 1966). The plankton is commonly a mixture of freshwater and brackish-water species, and the benthos is typically dominated by a few chironomid species. Several of the larger lakes are highly productive, particularly of rough fish. Shallower areas either support no fish, because of seasonal freezing, or primarily contain bullheads. Certain aquatic species appear to thrive in these environments (Wilson 1958). The amphipod *Hyalella azteca*, for example, is often found in abundance in the vegetated littoral areas (Eddy 1966).

Direct anthropogenic disturbances are largely restricted to the southern and western portions of this WRR where municipal waste additions and agricultural runoff have been largely responsible for biotic changes (Geraghty et al. 1973). Increased eutrophication and pesticide accumulation in aquatic species have been documented for these areas (Eddy 1966). Although the northeast portion of Minnesota has largely remained pristine, the preponderance of very soft waters makes this area exceedingly susceptible to acid rain effects and toxic metal effects.

MISSOURI BASIN

Hydrology, Water Quality, and Water Use

The Missouri Basin Water Resource Region, with an area of about 1,330,000 km^2 (515,000 sq miles) (U.S. Geological Survey 1977), is the largest region in the conterminous United States. The region includes the Missouri River from its headwaters in western Montana to near its confluence with the Mississippi River. (The most downstream reach of the Missouri River, from the mouth of the Osage River in central Missouri, to its confluence with the Mississippi River near St. Louis, Missouri, is considered part of the Upper Mississippi region.) The region also includes major tributary systems (Yellowstone and Platte river systems) and other prominent rivers such as the James, Big Horn, Powder, Little Missouri, Belle Fourche, Niobrara, and Republican rivers. The region includes some of the largest reservoirs (in terms of storage capacity) in the country, such as the Oahe,

Garrison, Fort Peck, and Fort Randall reservoirs on the Missouri River mainstem. Other major reservoirs are located on the Missouri River (Canyon Ferry Reservoir) and on tributary rivers, such as Tiber Reservoir (Marias River), Yellowtail Reservoir (Big Horn River), Seminoe and Pathfinder reservoirs and McConaughy Lake (North Platte River), Milford Reservoir (Republican River), Tuttle Creek Reservoir (Big Blue River), and Kaysinger Bluff Reservoir and Lake of the Ozarks (Osage River). Large natural lakes are not prominent features of the region, but scattered smaller lakes do exist.

Surface waters of the Missouri Basin Water Resource Region vary from soft (<60 mg/liter hardness as $CaCO_3$) in parts of northeastern Colorado to very hard in other parts of the region. In general, the hardest waters (>240 mg/liter) are found in the central and lower parts of the basin, from southeastern Montana and northeastern Wyoming through western and southern South Dakota and throughout large parts of Nebraska as far down the mainstem Missouri River as Kansas City, Missouri. Very hard waters (180 to 240 mg/liter) tend to be found in other parts of the lower and central basin; moderately hard to hard waters (60 to 180 mg/liter) are typically found in the upper Missouri River basin.

Levels of TDS are generally high (>350 mg/liter) in the region; low to moderate levels (<120 to 350 mg/liter) are found in headwater areas in Montana, Wyoming, and Colorado. Moderate concentrations (120 to 350 mg/liter) of TDS are found in parts of the Platte River basin in Nebraska, but high levels are typical of the mainstem Platte, North Platte, and South Platte rivers. Salinities may exceed 1000 mg/liter in parts of the Powder, Big Horn, James, and North and South Platte rivers and in a large area covering southwestern North Dakota and western South Dakota. Total suspended solids levels are typically low (<270 mg/liter) only in the Missouri River mainstem, in Fort Peck Reservoir, from near Garrison Reservoir downstream to the mouth of the Missouri River, the James River in and below Jamestown Reservoir, and in some headwater areas in Montana, Wyoming, and Colorado. Otherwise, moderate (270 to 1900 mg/liter) to high (>1900 mg/liter) concentrations are prevalent (Geraghty et al. 1973).

The most serious water quality problems in the Missouri Basin region occur as a result of nonpoint sources, some of natural origin. Headwaters, especially of snowmelt origin in Wyoming and Montana, are of excellent quality, but waters are degraded downstream (Water Resources Council 1974). Irrigation returns, crop and pastureland erosion, and feedlot wastes contribute sediments, bacteria, nutrients, pesticides, and salts. Dewatering of streams exacerbates these problems. Point sources, including municipal and industrial discharges, have been important in populated and developed areas, but such sources are seen as less severe and more easily controlled than nonpoint sources (U.S. Environmental Protection

Agency 1977b). Elevated levels of cadmium have been reported
in waters of the Missouri River basin (Geraghty et al. 1973).
Natural sources account for about 13% of degraded stream
miles in Montana; natural sources of suspended and/or dissolved solids are also identified as important in North
Dakota, Kansas, and Missouri. Other water quality problems
have been attributed to oil field wastes in Wyoming and mine
drainage in Kansas. Eutrophication of sections of the North
Platte and Laramie rivers in Wyoming has been reported.
Increased population demands, industrial growth, and mining
have been cited as threats to regional water quality (U.S.
Environmental Protection Agency 1977b).

The estimated amount of groundwater in storage with TDS
concentrations less than 3000 mg/liter is 4.2×10^{12} m^3
(3.4×10^9 acre-ft), or 64 times the average annual flow of
the Missouri River at Hermann, Missouri (Table XV). Groundwater occurs in three types of aquifers: unconsolidated,
semiconsolidated, and sedimentary rock.

Table XV. Estimated groundwater in storage
in the Missouri Basin region

Area	Groundwater in storage $[10^9$ m^3 $(10^6$ acre-ft)$]$
Northeastern Montana	0.5 (0.041)
Western Montana	0.1 (0.08)
Central Montana	0.2 (0.16)
Southeastern Montana and northeastern Wyoming	623.0 (505)
Western North Dakota and western South Dakota	566.4 (459)
Eastern North Dakota and eastern South Dakota	368.2 (299)
Southeastern Wyoming and northeastern Colorado	56.6 (45.9)
Northern and central Nebraska	1529.3 (1240)
Western Iowa, eastern Nebraska, and northeastern Kansas	93.5 (75.8)
Northern Kansas and southern Nebraska	566.4 (459)
Northwestern Missouri	424.8 (345)
Total	4248 (3450)

Source: Adapted from Taylor (1978).

Unconsolidated groundwater reservoirs include valley-
fill alluvium, alluvium and dune sand, and glacial and basin-
fill deposits (Table XVI). Gravel, sand, silt and clay
deposits that are hydraulically connected to major streams
comprise the valley-fill alluvial aquifers. Aquifer thickness averages 15 m (50 ft) but can attain a maximum of 73 m
(240 ft) along the South Platte River in Colorado. Well
yields are good; a large capacity well can produce up to
252 liters/sec (4000 gpm). Groundwater quality ranges from
approximately 100 to 4000 mg/liter TDS; however, the concentrations generally do not exceed 1000 mg/liter. Recharge
occurs through precipitation, irrigation water returns,
leakage from adjacent aquifers, and in losing stream reaches.
Discharge takes place primarily through gaining stream
reaches, evapotranspiration, well pumpage, and leakage to
adjacent aquifers (Taylor 1978).

An areally extensive alluvial aquifer hydraulically-
connected with valley-fill, dune-sand, and the underlying
Ogallala aquifers occurs in northeastern Kansas and southcentral Nebraska. Aquifer thickness is on the order of
perhaps 60 to 75 m (several hundred feet). Large capacity
wells used for irrigation yield up to 126 liters/sec
(2000 gpm). The TDS concentration seldom exceeds 500 mg/liter.
An extensive dune-sand aquifer is located in central Nebraska
and southern South Dakota; its maximum thickness is approximately 70 m (several hundred feet). The concentration of
dissolved solids is only about several hundred milligrams

Table XVI. Summary of water supplies in unconsolidated
aquifers of the Missouri Basin region

Source of supply	Location	Yield to wells [liters/sec (gpm)]
Valley-fill alluvial aquifers	Montana, Wyoming, North Dakota, South Dakota, Colorado, Nebraska, Minnesota, Iowa, Kansas, Missouri	0.63–252 (10–4000)
Alluvial aquifer, dune-sand aquifer	Nebraska, Kansas	3.15–126 (50–2000)
Glacial-deposit aquifers	Alberta, Saskatchewan, Montana, North Dakota, South Dakota, Nebraska, Minnesota, Iowa, Kansas, Missouri	0.63–126 (10–2000)
Basin-fill aquifers	Southwestern Montana	0.95–63 (15–1000)

Source: Adapted from Taylor (1978).

per liter. Recharge occurs primarily through precipitation; most of the discharge takes place through evapotranspiration and gaining reaches of streams (Taylor 1978).

Glacial aquifers occur over a width of as much as 322 km (200 miles). The maximum thickness attained is about 244 m (800 ft). Deposits consist of sand and gravel distributed in buried channels and outwash. Well yields are highly variable, but can attain 126 liters/sec (2000 gpm). Water quality is also variable, ranging from 300 to 30,000 mg/liter (Taylor 1978). Recharge occurs through precipitation, leakage from underlying consolidated aquifers, and losing streams. Discharge takes place through streams, evapotranspiration, underlying aquifers, or wells.

Basin-fill aquifers are found in southwestern Montana. These tertiary deposits attain a maximum thickness of approximately 1830 m (6000 ft). Well yields, although generally less than those typical of the other unconsolidated aquifers, can reach 63 liters/sec (1000 gpm). Concentration of TDS seldom exceeds 1000 mg/liter (Taylor 1978). The basin-fill aquifers are recharged by irrigation returns, precipitation, and losing streams. Groundwater is discharged to streams, wells, and the atmosphere (Taylor 1978).

Semiconsolidated groundwater reservoirs include the Ogallala, Arikaree, and Brule aquifers (Table XVII). The Ogallala Formation has a maximum thickness of 122 m (400 ft); it is comprised of semiconsolidated sand, gravel, silt, clay, and caliche. Large-capacity wells produce up to 227 liters/sec (3600 gpm); an average well yield is approximately 63 liters/sec (1000 gpm). Dissolved-solids concentrations range between 200 and 600 mg/liter. Recharge occurs from precipitation and leakage from adjacent aquifers. Discharge takes place mainly to wells and streams. Discharge is currently exceeding recharge, and water levels are dropping proportionately. The annual decline in Colorado is averaging 0.9 m (3 ft); in Kansas the annual decline is 0.2 m (0.7 ft) (Taylor 1978).

Table XVII. Summary of water supplies in semiconsolidated aquifers of the Missouri Basin region

Source of supply	Location	Yield to wells [liters/sec (gpm)]
Ogallala aquifer	Wyoming, South Dakota, Colorado	0.63-227 (10-3600)
Arikaree aquifer	Wyoming, South Dakota, Colorado, Nebraska	0.63-126 (10-2000)
Brule aquifer	Wyoming, South Dakota, Colorado, Nebraska	0.63-126 (10-2000)

Source: Adapted from Taylor (1978).

The semiconsolidated deposits of the Arikaree Formation underlie the Ogalalla and consist of sandstone, siltstone, and claystone, with a maximum thickness of 183 m (600 ft). Wells produce up to 126 liters/sec (2000 gpm) of high quality groundwater (several hundred milligrams per liter TDS). Precipitation and irrigation return flows constitute most of the recharge to the Arikaree aquifer; discharge is to streams and wells. The Arikaree water table is declining in northwestern Nebraska as a result of groundwater mining (Taylor 1978).

The Brule Formation, which is comprised of siltstone and fractured sandy clay, underlies the Arikaree Formation. The maximum thickness attained by the Brule sediments is 137 m (450 ft). Wells yield up to 126 liters/sec (2000 gpm) because of the highly fractured nature of the aquifer material. Dissolved solids concentrations seldom exceed several hundred mg/liter. The Brule aquifer is recharged by precipitation and influent streamflow; discharge is accommodated by flow to streams and wells. Minor declines in the water level have occurred in Wyoming and Nebraska (Taylor 1978).

Consolidated rock aquifers are composed of sandstone, limestone, and dolomite. Sandstone aquifers occur throughout the Missouri Basin region, with the exception of most of Nebraska, Kansas, and Missouri, and parts of Montana, Wyoming, and Colorado (Table XVIII). The principal sandstone aquifers are the Dakota Sandstone, Inyan Kara Group, and Sundance, Minnelusa, and Deadwood formations, and the Jordan Sandstone (Taylor 1978).

The Dakota Sandstone crops out over central Kansas, western Iowa, and eastern Nebraska. Its maximum thickness is 213.4 m (700 ft). Well production ranges up to 94.6 liters/ sec (1500 gpm); flowing wells are common. Levels of TDS vary, with a range of 300 to 30,000 mg/liter; saline zones occur within the aquifer (Taylor 1978). Recharge is mainly from precipitation on outcrop areas and interaquifer flow; discharge is to streams and other aquifers.

The Inyan Kara and Sundance aquifers immediately underlie the Dakota aquifer in South Dakota. Well yields are reportedly high, but TDS concentrations vary from 1000 to more than 10,000 mg/liter (Taylor 1978).

The Minnelusa and Deadwood aquifers also underlie the Dakota aquifer. Water quality has not yet been adequately defined; however, well yields of at least 252.4 liters/sec (4000 gpm) have been obtained (Taylor 1978).

The Jordan Sandstone of Iowa is in hydraulic connection with the overlying Prairie du Chien Formation (a dolomite); these produce as a unit. The thickness of the aquifer is from 91.4 to 182.9 m (300 to 600 ft). Wells yield up to 63.1 liters/sec (1000 gpm). The concentration of dissolved solids varies from 1500 to 2000 mg/liter (Taylor 1978).

Table XVIII. Summary of water supplies in sandstone aquifers of the Missouri Basin region

Source of supply	Location	Yield to wells [liters/sec (gpm)]
Virgelle aquifer, Milk River aquifer	Montana, Alberta	0.32–15.8 (5–250)
Blood Reserve, Judith River, and Milk River aquifers	Alberta	0.32–7.6 (5–120)
Judith Basin	Montana	0.32–18.9 (5–300)
Hogeland Basin	Saskatchewan	0.63–12.6 (10–200)
Crazy Mountains and Bull Mountains Basins	Montana	
Bighorn Basin	Wyoming, Montana	0.32–126.2 (5–2000)
Wind River Basin	Wyoming	0.32–69.4 (5–1100)
Laramie, Shirley, and Hanna Basins and Saratoga Valley	Wyoming	0.32–63.1 (5–1000)
Powder River Basin	Montana, Wyoming	0.32–75.7 (5–1200)
Williston Basin	Montana, North Dakota, South Dakota	0.32–44.2 (5–700)
Denver Basin	Colorado	0.63–18.9 (10–300)
Dakota aquifer	Nebraska, Iowa, Kansas, Minnesota	0.63–94.6 (5–1500)
Dakota, Inyan Kara, Sundance, Minnelusa, and Deadwood aquifers	North Dakota, South Dakota	0.63–252.4 (5–4000)
Prairie du Chien and Jordan aquifers	Iowa	6.3–63.1 (100–1000)

Source: Adapted from Taylor (1978).

The carbonate aquifers of the Missouri Basin region produce from solution openings, joints, and fractures. The primary limestone and dolomite aquifers are the Madison Limestone, Bighorn Dolomite-Red River Formation-Whitewood Dolomite, Roubidoux Formation, and Gasconade, Eminence, and Potosi dolomites (Table XIX).

The Madison Limestone underlies portions of the states of Montana, Wyoming, North Dakota, South Dakota, Nebraska, and Colorado. It attains its maximum thickness in the central Williston Basin of North Dakota; here, its thickness exceeds 610 m (2000 ft). Wells produce up to several hundred liters per second of generally good quality (300 mg/liter) groundwater; however, with increasing depth and temperature, the TDS concentration may reach 350,000 mg/liter (Taylor 1978). The Madison aquifer is recharged primarily by precipitation and streamflow in outcrop areas. The aquifer discharges to streams, springs, wells, and other aquifers.

Table XIX. Summary of water supplies in limestone and
dolomite aquifers of the Missouri Basin region

Source of supply	Location	Yield to wells [liters/sec (gpm)]
Madison aquifer	Alberta, Saskatchewan, Montana, Wyoming, North Dakota, South Dakota, Nebraska, Colorado	1.3–567.8 (20–9000)
Limestone and sandstone aquifers in rocks of Permian and Pennsylvanian age	Iowa, Nebraska, Kansas, Missouri	0.19–31.6 (3–500)
Bighorn, Red River, and Whitewood aquifer	Montana, Wyoming, North Dakota, South Dakota	1.26–18.9 (20–300)
Roubidoux, Gasconade, Eminence, and Potosi aquifers	Missouri, Kansas	1.26–126.2 (20–2000)

Source: Adapted from Taylor (1978).

The Bighorn–Red River–Whitewood aquifer, which underlies
the Madison aquifer in the Williston Basin and in portions of
Montana and Wyoming, has developed in approximately equivalent
dolomite formations. The aquifer attains a thickness of
168 m (550 ft). Well yields range up to perhaps several tens
of liters per second. The recharge/discharge characteristics
of the aquifer have not been defined, but there seems to be
some interaction with the Madison aquifer (Taylor 1978).

The Roubidoux, Gasconade, Eminence, and Potosi series of
dolomite aquifers comprise the principal aquifers of the
Ozarks. The aquifers are very transmissive; groundwater may
flow 1.6 km (1 mile) per day. Maximum well yield is
126 liters/sec (2000 gpm). Bennett Spring, which issues
from the aquifers, has a discharge of approximately
4250 liters/sec (67,361 gpm). The concentration of TDS
seldom exceeds 350 mg/liter (Taylor 1978). Aquifer recharge
is from precipitation and streamflow. The aquifers discharge
primarily to streams.

In the Missouri Basin Water Resource Region, average
annual runoff is equivalent to 200×10^6 m^3/day (54 Bgd).
Of the total off-channel withdrawal of 130×10^6 m^3/day
(35 Bgd) in 1975, surface freshwater withdrawals accounted
for about 95×10^6 m^3/day (25 Bgd), and fresh groundwater
accounted for the remainder. About 43% of freshwater withdrawn was consumed. Irrigation accounts for the majority of
water withdrawals and freshwater consumption (Table I).
Irrigation also is responsible for almost 35×10^6 m^3/day
(9 Bgd) of the total groundwater withdrawals. The Missouri

Basin region uses about 570×10^6 m^3/day (150 Bgd) for hydro-electric power, with the region ranking seventh in the nation in this respect (U.S. Geological Survey 1977). The Montana, North Dakota, and Wyoming portions of the Missouri Basin Water Resource Region have been identified by the Water Resources Council as being among the nation's most critical energy-related water-problem areas. In particular, the extraction and use of coal resources in these areas is seen as increasing agricultural vs industrial competition for water. Development of additional storage capability and importation of water from the Upper Colorado basin are suggested as augmenting dependable water supplies in the region (Water Resources Council 1974).

Aquatic Ecology

Although this region is the largest in the country, it contains comparatively few natural lakes. Rivers, however, are a conspicuous aquatic resource, and most of the Missouri River is contained within the WRR.

The headwaters of the Missouri River in the mountains of Montana are quite different in character than the downstream waters. Near its origin the river flows over a substrate of rock and gravel, and a typical trout stream biota exists (Hynes 1972). However, most of the river lies within the Great Plains and Central Lowland provinces where the water is silt-laden, the substrate consists of soft sediments, and the biota is impoverished (U.S. Nuclear Regulatory Commission 1977).

The river has been greatly altered by human activity. Channelization and damming, for example, have greatly reduced the average surface area and have created stronger currents (with resultant bottom scour) and greater siltation and turbidity than existed a century ago (U.S. Nuclear Regulatory Commission 1977). Although the water is moderately hard to hard and nutrient levels are generally high, the river is relatively unproductive, largely because of the lack of favorable habitats (U.S. Nuclear Regulatory Commission 1977). The phytoplankton is dominated by diatoms and green algae; average densities are on the order of 10^7–10^8 cells per cubic meter. The development of greater densities is impeded by swift flows and high turbidities, and many of the organisms likely represent washout from upstream reservoirs. Copepods and rotifers dominate the zooplankton, but total numbers are low. The macroinvertebrates in the drift and in benthic areas primarily consist of caddis fly larvae, mayfly larvae, and midgefly larvae (U.S. Nuclear Regulatory Commission 1977). Channel catfish, buffalo fish, and carp are actively fished commercially. The most abundant fish species present include carp, freshwater drum, gizzard shad, river carpsucker, shortnose gar, paddlefish, crappies, and shiners (U.S. Nuclear Regulatory Commission 1977).

Many of the same organisms exist in the impounded areas of the river, but the current velocities and turbidities are lower, thus permitting the development of a more biologically productive system (Hynes 1972). Likewise, peripheral areas in these reservoirs provide some spawning habitat for other warmwater fishes, such as centrarchids. In areas where nutrient inputs are large, substantial blue-green algal blooms can develop (Pennak 1949).

The smaller prairie streams within this WRR are similar to those in the Arkansas–White–Red WRR and the small prairie lakes are like those in the Souris–Red–Rainy WRR (see sections on these regions).

The extreme western portion of this region contains high altitude areas with mountain streams and lakes. These are described in the sections on the Texas Gulf and Upper Colorado regions.

In addition to stream alteration via damming and channel-ization, major anthropogenic disturbances that have disrupted the natural biota in this region include siltation, pesticide and fertilizer inputs from agricultural operations, and municipal waste additions. In static, nonturbid water, excessive eutrophication effects are common, because of high nutrient inputs (Carlander, Campbell, and Irwin 1966). Several polluted farm ponds and natural waters have been responsible for the deaths of wildlife and livestock when blooms of toxic blue-green algae developed (Hynes 1971). Mine drainage effects (particularly from coal strip mining) occur and are severe in some areas; however, the aridity of the areas in which mining takes place limits runoff (Hynes 1972; Carlander, Campbell, and Irwin 1966).

UPPER COLORADO

Hydrology, Water Quality, and Water Use

The Upper Colorado Water Resource Region, with an area of about 285,000 km^2 (110,000 sq miles) (U.S. Geological Survey 1977), includes the headwaters of the Colorado River and major tributary systems, such as the Green and San Juan river systems. Lees Ferry, Arizona, just downstream from the Glen Canyon Dam, is taken as the dividing point between the Upper and Lower Colorado River basins. Large natural lakes are not important features of the region. However, reservoirs are widespread, ranging from Lake Powell, the impound-ment of Glen Canyon Dam (the second largest reservoir in the United States in terms of storage capacity) (Geraghty et al. 1973), to the many small reservoirs, such as Strawberry and

Big Sandy reservoirs. Other major reservoirs include Flaming
Gorge Reservoir on the Green River and Navajo Reservoir on
the San Juan River.

Surface waters of the Upper Colorado Water Resource
Region are of moderate hardness (60 to 120 mg/liter hardness
as $CaCO_3$) only in the headwaters of the Colorado Rockies;
elsewhere, waters are hard to very hard (120 to 240 mg/liter
and above). The hardest waters (>240 mg/liter) are found
along the Colorado River mainstem in southeastern Utah.
Total dissolved solids levels follow a similar pattern: low
to moderate levels (<120 to 350 mg/liter) in the headwaters
of the Colorado Rockies, with high levels (>350 mg/liter)
elsewhere. The mainstem Green and San Juan rivers and the
Colorado River in extreme western Colorado and downstream are
typically saline (TDS >1000 mg/liter). Total suspended
solids levels are low (<270 mg/liter) in the Colorado Rockies
headwaters, in the Green River mainstem in northeastern Utah
and southwestern Wyoming, in Lake Powell, and in the upper
San Juan River mainstem in New Mexico. Intermediate levels
(270 to 1900 mg/liter) are found in other parts of western
Colorado. Otherwise, high levels (>1900 mg/liter) are
prevalent (Geraghty et al. 1973).

Regional water quality problems include high natural and
man-made levels of suspended and dissolved solids. Natural
sources include diffuse sources (such as surface runoff),
which increase sediment and salinity loads, as well as point
sources (springs). Heavy surface-water consumption reduces
the dilution capabilities of affected streams, exacerbating
the problem. Irrigation return flows have also adversely
affected surface-water quality (U.S. Environmental Protection
Agency 1977b). In the Green River basin, for example, while
waste water treatment and low population densities limit
municipal and industrial pollution sources, irrigated lands
yield an estimated 4500 to 13,450 kg of salt per hectare per
year (2 to 6 tons of salt per acre per year) to surface
waters via return flows (Wyoming Water Planning Program 1970).
Control of point sources such as municipal discharges is more
easily assured. For example, improvement of the San Juan
River in New Mexico was evident following improved waste
water treatment (U.S. Environmental Protection Agency 1977b).
Problems with coliform bacteria, biochemical oxygen demand
(BOD), dissolved solids, sulfate, and certain metals (lead,
cadmium, iron) have been reported from waters of the Colorado
River system in Utah (U.S. Environmental Protection Agency
1977b).

The maximum volume of recoverable groundwater in storage
in the upper 30.5 m (100 ft) of waterbearing material in the
Upper Colorado region is estimated to be 6.2×10^9 m^3
(50×10^6 acre-ft) (Price and Arnow 1974). Eighty-five
percent of this amount occurs in sedimentary rocks; 5% is
stored in unconsolidated material. The total amount of

recoverable groundwater present in the entire aquifer sequences cannot easily be estimated but is thought to be many times the quantity stored within the upper 30.5 m (100 ft) of saturated rock material. Groundwater is stored in five general types of hydrogeologic units (Table XX). The following aquifer data are from Price and Arnow (1974).

The unconsolidated deposits consist of gravel, sand, and clay. They occur primarily along the larger stream valleys. Depths to water are generally less than 15 m (50 ft). Yields to wells vary between 0.3 to more than 31.6 liters/sec (5 to more than 500 gpm).

The volcanic rocks consist primarily of lava flows, with some associated pyroclastic and intrusive igneous rocks. These aquifer units occur mostly along the east and west-central borders of the region. Well yields usually range from 0.3 to 3.16 liters/sec (5 to 50 gpm) but can locally range from 3.16 to 31.6 liters/sec (50 to more than 500 gpm).

The sedimentary rock aquifers are divided into two classes on the basis of origin: nonmarine and marine. Both types occur throughout the region. Depths to water are from several tens to more than 304.8 m (several hundred to more than 1000 ft). The nonmarine aquifers are more permeable and generally yield from 0.3 to 3.16 liters/sec (5 to 50 gpm) to wells. The aquifers of marine origin yield less than 0.6 liters/sec (10 gpm) to wells.

Granite, schist, gneiss, and quartzite comprise the igneous and metamorphic rocks. Their occurrence is primarily restricted to the eastern boundary, the northernmost, and the northcentral portions of the region. Well yields rarely exceed 0.6 liters/sec (10 gpm).

The quality of groundwater in the region is good; a very few restricted areas have TDS concentrations exceeding 3000 mg/liter. Shallow aquifers at higher elevations [above 2133.6 m (7000 ft)] provide water with less than 1000 mg/liter TDS, as do the sandstone (Navajo and Dakota Sandstones) and limestone (Madison Limestone and Morgan Formation) aquifers at lower elevations. Shale and siltstone formations usually contain saline groundwater.

Four percent [4.9×10^9 m^3 (4×10^6 acre-ft)] of the average annual precipitation in the region [1.2×10^{11} m^3 (95×10^6 acre-ft)] enters the groundwater reservoirs as recharge. This recharge occurs through infiltration of precipitation, irrigation water, and influent streams. Discharge takes place through springs, areas of phreatophyte growth, and gaining reaches of (effluent) streams.

Groundwater withdrawals in 1970 totalled 1.5×10^8 m^3 (0.12×10^6 acre-ft); consumptive use totalled 7.4×10^7 m^3 (0.06×10^6 acre-ft). This is approximately 2% of the total volume of water withdrawn [7.0×10^9 m^3 (5.7×10^6 acre-ft)] and consumed [4.4×10^9 m^3 (3.6×10^6 acre-ft)] that same year. Of the total groundwater withdrawn, irrigation

Table XX. Estimated recoverable groundwater in storage of the Upper Colorado region

Geohydrologic unit	Rock type	Area [10^7 m^2 (acres × 10^3)]	Saturated thickness [m (ft)]	Estimated amount of water in storage [10^7 m^3 (10^3 acre-ft)]	
				Minimum	Maximum
1	Unconsolidated deposits	324 (800)	15 (50)	247 (2,000)	740 (6,000)
2	Volcanic rocks	890 (2,200)	31 (100)	543 (4,400)	1,360 (11,000)
3	Sedimentary rocks	16,200 (40,000)	31 (100)	4,970 (40,300)	9,940 (80,600)
4	Sedimentary rocks	9,830 (24,300)	31 (100)	604 (4,900)	2,100 (17,000)
5	Igneous and metamorphic rocks	2,060 (5,100)	31 (100)	0 (0)	185 (1,500)
Total (rounded)				6,170 (50,000)	14,200 (115,000)

Source: Adapted from Price and Arnow (1974).

accounted for 46%, public supply for 27%, self-supplied
industry for 15%, and domestic and stock for 12%. The amount
of groundwater consumption can be attributed to irrigation
is 52%; to public supply, 25%; to domestic and stock, 17%;
and to self-supplied industry, 6% (Price and Arnow 1974).

In the Upper Colorado Water Resource Region, average
annual runoff is equivalent to 49×10^6 m^3/day (13 Bgd). Of
the total off-channel water withdrawal of 16×10^6 m^3/day
(4.1 Bgd) in 1975, surface freshwater withdrawals accounted
for 15×10^6 m^3/day (3.9 Bgd). Fresh groundwater was the
other major contributor. About 41% of freshwater withdrawn
was consumed. Irrigation accounts for the majority of water
withdrawals and freshwater consumption (Table I). The Upper
Colorado region uses about 49×10^6 m^3/day (13 Bgd) for
hydroelectric power, with the region ranking ahead of only
four other regions in the conterminous United States in this
respect (U.S. Geological Survey 1977).

The Upper Colorado Water Resource Region has been
categorized by the Water Resources Council as being among the
nation's most critical energy-related water-supply problem
areas. Because of international agreement with Mexico,
interstate compact, and Federal and state laws, not all
runoff in the region is available for use within the region.
Although estimates of water availability in the region (based
on required deliveries to Mexico and the Lower Colorado basin
and on current use rates) indicate that about 8×10^6 m^3/day
(2 Bgd) of water supply are not being utilized in the region,
not all of this amount is available. In addition to environmental water needs (water quality, fish and wildlife, and
aesthetics), uneven geographic distribution of supplies and
demands and variable annual runoff cause water-use problems.
Water supplies have been over-appropriated in some areas,
especially in Colorado and Utah, with existing water rights
exceeding current use and estimated resources. Competition
between agricultural and industrial demands is seen as a
result. Development of more surface storage, interim use of
groundwater, greater efficiency in water use, and resolution
of water rights and allocations are seen as crucial to the
satisfaction of anticipated water demands in the region
(Water Resources Council 1974).

Aquatic Ecology of the Central Western United States

Prominent physiographic features that largely determine
the occurrence of particular biota and habitats in this area
include mountains, plateaus, and low, arid basins.

The mountainous areas contain hundreds of small natural
lakes. Many of these have a constant flow of water through
them, whereas others lack any permanent outflow. The former
types generally display very low productivities, but the

80

latter ones are frequently moderately productive, because of the gradual buildup of nutrients in the closed basins (Juday 1907). Those lakes found at the highest altitudes are designated alpine lakes; they usually occur in rock and gravel basins and have high inflow and outflow rates. The biota of these lakes is typically depauperate because of the combination of a number of stresses imposed on the system, that is, long periods of time with ice cover, the addition of glacial "milk" (rock flour), and low temperatures (Pennak 1966). The phytoplankton densities are commonly less than 100,000/liter, and the benthos generally consists of a sparse assemblage of dipteran larvae and little else. Fish are often absent from these lakes, but endemic copepod species are common (Pennak 1966).

Montane lakes occur at lower altitudes and are generally more fertile. The benthos is often very diverse, and rooted aquatic plants are common. Blue-green algal blooms and summer bottom oxygen depletion occur in the more eutrophic lakes. These water bodies are commonly found in spruce-fir forests or alpine meadows (Pennak 1966).

Few lakes occur in the foothills region of the mountains, but many of the larger streams have been dammed to create reservoirs. In general, these impoundments resemble those described for the Texas Gulf WRR (Pennak 1949).

The more arid portions of the region contain numerous saline lakes and playas. The saline lakes lack a diverse biota but typically contain large standing crops of a few species. Particularly common are the brine shrimp (*Artemia salina*) and the algae *Stichococcus bacillaris* and *Dunaliella* sp. (Edmonson 1966). The Salton Sea, a large saline lake that receives considerable fertilizer runoff, periodically contains large densities of diatoms and dinoflagellates, as well as an assemblage of rotifers, copepods, and a few marine species. The benthic invertebrates of these lakes are largely restricted to a few species of dipteran larvae (Edmonson 1966).

The rivers that occur in the more arid portions of this area are similar to those described for the Texas Gulf region. Many of the smaller streams are ephemeral and do not support a stable biota (Edmonson 1966).

Increased siltation and turbidity in surface waters caused by land disturbances are major problems in this area. Another factor that has contributed greatly to habitat alteration and destruction is the extensive channelization and impounding that has occurred (Minckley 1973; Edmonson 1966). Parts of the area have experienced extreme eutrophication from fertilizer runoff, and biotic accumulation and/or direct toxic effects of pesticides are persistent problems in agricultural regions (Edmonson 1966).

Hydrology, Water Quality, and Water Use

The Lower Colorado Water Resource Region, with an area of about 355,000 km^2 (137,000 sq miles) (U.S. Geological Survey 1977), includes the Colorado River and its tributary systems below Lees Ferry, Arizona, just downstream from Glen Canyon Dam. Major tributary systems include the Virgin River, Little Colorado River, Verde-Salt-Gila rivers, and the Bill Williams River. The region does not contain large natural lakes, but the Lower Colorado system is heavily impounded.

Lake Mead, behind Hoover Dam on the Colorado River mainstem, has the largest storage capacity of U.S. reservoirs (Geraghty et al. 1973). Other major reservoirs are located on the Colorado River itself (Lake Mohave and Havasu Lake) and on the Gila (Painted Rock and San Carlos reservoirs), Salt (Roosevelt Reservoir), and Bill Williams (Alamo Reservoir) rivers.

Surface waters of the Lower Colorado Water Resource Region are generally hard (120 to 180 mg/liter as $CaCO_3$) to very hard (180 to 240 mg/liter) with the hardest waters (>240 mg/liter) found in the Virgin River and Colorado River mainstem. Total dissolved solids levels are generally high (>350 mg/liter), with waters of saline levels (>1000 mg/liter) found along the Virgin, Little Colorado, Salt, and upper Gila rivers. Total suspended solids concentrations are low (<270 mg/liter) where affected by impoundment (along the Colorado River mainstem from below Lake Powell to the confluence of the Little Colorado River and in and downstream from Lake Mead and in portions of the Gila and Salt rivers), but, otherwise, concentrations are generally high (>1900 mg/liter) (Geraghty et al. 1973).

Regional water quality problems are dominated by nonpoint-source (in some cases natural) discharges. Agricultural and rural sources, including runoff and irrigation return flows, contribute nutrients, sediments, salts, and bacteria (U.S. Environmental Protection Agency 1977b). The overall increasing salinity in a downstream direction (mostly natural but accelerated by man's activities) is a major regional problem. The salinization of soils and groundwater is intimately related (Water Resources Council 1974). Metals contamination from mineralized areas, particularly abandoned mining operations, has been cited as a problem in Arizona (U.S. Environmental Protection Agency 1977b). Elevated levels of cadmium and mercury have been reported from the Gila River system (Geraghty et al. 1973). Eutrophication is reported to be a problem in Arizona (U.S. Environmental Protection Agency 1977b).

The principal aquifers of the Lower Colorado region are unconsolidated sediments and consolidated sedimentary rocks. Well yields range from very low to moderately high.

The southern and western portions of the region consist of alternating fault-block basins and mountain ranges. The aquifers are almost entirely unconsolidated valley fill material (alluvial fan deposits) eroded from adjacent mountains. The sediments are primarily sand and gravel interbedded with silts and clays. Thicknesses vary between tens of meters to more than 1000 m (Walton 1970). The highest yields are produced from wells located at the mouths of mountain stream canyons.

The northeastern portion of the region is underlain by consolidated sedimentary rocks, such as sandstones, shales, and limestones. Some alluvial material is found in the stream valleys. The average well yields are very low; however, the Coconino Sandstone of Permian age is an excellent aquifer. Wells producing from the Coconino yield several tens of liters per second (hundreds of gallons per minute) (Walton 1970).

In the Lower Colorado Water Resource Region, average annual runoff is equivalent to about 12×10^6 m^3/day (3.2 Bgd) (U.S. Geological Survey 1977). By law, the Upper Colorado basin must deliver an average of 25×10^6 m^3/day (7.5×10^6 acre-ft/year) to the Lower Colorado basin at Lees Ferry, Colorado, calculated on a ten-year averaging interval (Hutchins 1977). However, not all this Colorado River water is available for use within the Lower Colorado Water Resource Region; it has been estimated that about 11×10^6 m^3/day (2.8 Bgd) is available to the region (Water Resources Council 1974).

Of the total off-channel water withdrawals of about 32×10^6 m^3/day (8.5 Bgd) in 1975, surface freshwater withdrawals accounted for about 13×10^6 m^3/day (3.5 Bgd), while fresh groundwater contributed the remainder. Of the freshwater withdrawals, about 74% was consumed. The Lower Colorado Water Resource Region uses about 91×10^6 m^3/day (24 Bgd) for hydroelectric power, ranking ahead of six other regions in the conterminous United States in this respect (U.S. Geological Survey 1977). Irrigation is the dominant water user and consumer of freshwater (Table I). Irrigation also accounts for almost 90% of groundwater use (U.S. Geological Survey 1977).

The Water Resources Council has identified the Lower Colorado Water Resource Region as one of the nation's most critical energy-related water-supply problem areas. The Colorado River system as a whole is seen as unable to continually supply demands in the Upper and Lower Colorado basins and Mexico as per treaty obligations (Water Resources Council 1974). For example, California is entitled to 5.4×10^9 m^3/year of Colorado River water, according to a 1964 U.S. Supreme

Court decree (Imperial Irrigation District 1977). The Rio
Grande, Colorado, and Tijuana Treaty of 1944 guarantees
Mexico 1.9×10^9 m^3 (1.5×10^6 acre-ft) of Colorado River
water annually (Hutchins 1977). Use of groundwater currently
exceeds recharge; overdraft has exceeded an average of
8×10^6 m^3/day (2 Bgd) in the Gila River basin. The importation of water from areas such as Alaska, Canada, or the
Pacific Northwest has been proposed. The conclusion is that
energy-related water demands will compete with irrigation and
municipal and industrial needs. Water development projects
such as the Central Arizona Project, which will allow
Colorado River water to be used by the Tucson-Phoenix,
Arizona, area and enroute agricultural lands, are expected to
alleviate distributional problems (Water Resources Council
1974).

Aquatic Ecology

The discussion of biotic resources in the Texas Gulf
WRR section is applicable to the Lower Colorado WRR.

GREAT BASIN

Hydrology, Water Quality, and Water Use

The Great Basin Water Resource Region, with an area of
about 480,000 km^2 (185,000 sq miles) (U.S. Geological Survey
1977), includes large portions of Nevada and Utah and smaller
portions of adjacent states. Drainage is interior (land-
locked). Large portions of the region are devoid of permanent surface waters. Saline lakes, ranging in size up to
that of the Great Salt Lake, are widely distributed. Feeding
these lakes are streams and stream systems that range in size
from small ephemeral streams to large systems such as the
Humboldt, Truckee, and Bear rivers. Lake Tahoe, an excep-
tionally deep freshwater lake, is fed by snowmelt from the
Sierra Nevada Range. The region also contains many channels
and basins (washes and playas) that contain water only
ephemerally, as a result of precipitation events.
Surface waters of the Great Basin region tend to
increase in hardness in an eastward direction. Soft waters
(<60 mg/liter hardness as $CaCO_3$) are found in extreme western
Nevada; most of Nevada has moderately hard (60 to 120 mg/liter)
waters. The Carson Sink area and the eastern part of the
Great Basin have waters ranging from hard to very hard (120
to >240 mg/liter); the hardest waters occur in southwestern
Utah. Surface waters generally have high TDS concentrations
(>350 mg/liter), although moderate levels (120 to 350 mg/liter)
are found in the western and northern extremes of the basin

and in parts of southwestern Utah. The lowest concentrations
(<120 mg/liter) are typical only around Lake Tahoe. Saline
waters (>1000 mg/liter) are found in the Nevada-California
border area (Pyramid, Mono, and Walker lakes) and in Utah
(Great Salt Lake and Utah Lake). Stream TSS levels generally
range from low (<270 mg/liter) in the California-Nevada
border area and in the Great Salt Lake Basin to moderate (270
to 1900 mg/liter) in northern Nevada and parts of Utah to
high (>1900 mg/liter) elsewhere in the Great Basin (Geraghty
et al. 1973).

Surface water quality in the Great Basin is largely
limited by natural nonpoint sources of mineralization. While
municipal and industrial point-source discharges are of local
significance (as in the Lower Provo and Jordan rivers in
Utah), the diffuse salinity and sediment contribution of
agricultural practices is of greater regional significance
(U.S. Environmental Protection Agency 1977b). Cultural
eutrophication has been noted in the ultraoligotrophic Lake
Tahoe (Goldman 1972; U.S. Environmental Protection Agency
1977d).

The volume of recoverable groundwater in storage within
the upper 30 m (100 ft) of aquifers in the Great Basin region
is estimated to be 3.7×10^{11} m^3 (300×10^6 acre-ft). Total
groundwater storage within the entire aquifer thickness is
probably on the order of several trillion cubic meters
(several billion acre-ft). The regional groundwater reservoirs are of three basic types: unconsolidated alluvial
material, consolidated carbonate rocks, and consolidated
volcanic rocks. Large capacity wells average yields of
63 liters/sec (1000 gpm) (Eakin, Price, and Harrill 1976).

The unconsolidated alluvial aquifers are primarily
comprised of sand and gravel deposits; however, in the lowest
portions of valleys, silt and clay deposits are common. The
alluvial aquifers occur in the intermontane basins of the
region. The potentiometric surface is relatively close to
land surface in the lowest areas; depth to water may increase
several tens of meters towards the mountains. Aquifer thickness varies from site to site, often exceeding 300 m (1000 ft)
(Eakin, Price, and Harrill 1976).

The carbonate rock aquifers of the region are generally
highly permeable and occur in the central and northeastern
portions of the region. Large springs discharge water from
carbonate aquifers in eastern Nevada and western Utah. Many
perched groundwater bodies exist where the carbonates are
discontinuous or offset by faults.

Known volcanic rock aquifers of significant size occur
only in the southeastern and extreme northeastern portions of
the region. Large quantities of water can be obtained from
the fracture openings within the volcanics.

Groundwater quality varies from fresh (less than 1000 mg/liter TDS) to briny (more than 35,000 mg/liter TDS) across the region (Eakin, Price, and Harrill 1976). Freshwater is most commonly found at valley margins. Groundwater underlying small basins may be brackish; major deep valleys (sinks) are underlain by brackish to briny groundwater. Locally, thermal springs have water of poor quality.

The average annual precipitation in the Great Basin region is 1.1×10^{11} m^3 (88×10^6 acre-ft). Of this amount, 3 to 7% [4×10^9 to 7×10^9 m^3 (3×10^6 to 6×10^6 acre-ft)] recharges the groundwater reservoirs. Most recharge occurs through infiltration of precipitation in the mountains; some input from losing streams takes place. Groundwater discharge is through evapotranspiration, spring discharge, contribution to stream flow, and underflow to adjacent areas (Eakin, Price, and Harrill 1976).

In the Great Basin Water Resource Region, average annual runoff is equivalent to 28×10^6 m^3/day (7.5 Bgd). Of the total off-channel water withdrawals of 26×10^6 m^3/day (6.9 Bgd) in 1975, surface freshwater withdrawals accounted for about 20×10^6 m^3/day (5.4 Bgd) and fresh groundwater withdrawals about 5.3×10^6 m^3/day (1.4 Bgd). Of freshwater withdrawals, about 53% was consumed. The Great Basin region uses about 14×10^6 m^3/day (3.8 Bgd) for hydroelectric power, greater in this respect than only the Rio Grande and Souris–Red–Rainy regions in the conterminous United States. Irrigation accounts for the majority of total withdrawals and freshwater consumption (Table I). Irrigation also accounts for about 70% of fresh groundwater use (U.S. Geological Survey 1977).

The Water Resources Council has designated the Great Basin Water Resource Region as one of the nation's most critical energy-related water-supply problem areas. Water-related problems result from the aridity of the region and the expense of water development. Importation of water from the Colorado River Basin to the Wasatch Front area of the eastern part of the Great Basin by construction of the Bonneville Unit of the Central Utah Project is seen as partly alleviating problems. However, water supplies are still expected to be inadequate in the western part of the region around Reno, Nevada. Additional water demands are considered limiting to increased energy development (Water Resources Council 1974).

Aquatic Ecology

The discussion of biotic resources in the section on the Upper Colorado WRR is applicable to the Great Basin WRR.

CALIFORNIA

Hydrology, Water Quality, and Water Use

The California Water Resource Region, with an area of
about 310,000 km^2 (120,000 sq miles) (U.S. Geological Survey
1977), consists largely of the river systems (e.g.,
Sacramento, San Joaquin) that drain the western slopes of
the Sierra Nevada Mountains and flow toward San Francisco
Bay. Reservoirs, such as Shasta Lake and Folsom and Pine
Flat reservoirs, are located on headwaters of these streams.
Other river systems with drainage into the Pacific Ocean
include the Klamath, Russian, Salinas, Eel, Cuyama, and Santa
Clara rivers. Large natural lakes include Upper Klamath
Lake, Clear Lake, and Buena Vista Lake (at the terminus of
the Kern River). The Salton Sea, in the Imperial Valley of
southern California, was formed and is fed by water from the
lower Colorado River. Major aqueducts and canals (Hetch
Hetchy, Los Angeles, and San Diego—Colorado River aqueducts
and the All-American and Coachella canals) provide for water
transport to municipal, industrial, and agricultural demand
centers along the coast and in the Imperial Valley.
    Surface water hardness of the California WRR generally
increases southward and coastward from soft (<60 mg/liter
hardness as CaCO$_3$) in the Sierras and in northern California
to moderate (60 to 120 mg/liter) at the foothills of the
Sierras and along the northern coast to hard (120 to 180 mg/
liter) in the San Joaquin Valley and along the southern
coast. Very hard waters (180 to 240 mg/liter) are found
inland in southern California and along the coast in the Los
Angeles-San Diego area. Levels of TDS also generally increase
from lowest (<120 mg/liter) in the Sierras and in northern
California to higher (>350 mg/liter) in the southern interior
and coast. Saline waters (>1000 mg/liter) are found scattered
throughout southern California. The Salton Sea is the third
largest saline lake [907 km^2 (350 sq miles)] in the nation.
Stream TSS concentrations generally increase from low
(<270 mg/liter) in northern California and the Sierras to
moderate (270 to 1900 mg/liter) along most of the coast and
to high (>1900 mg/liter) near San Diego and in southern
interior California (Geraghty et al. 1973).
    Although overall surface water quality in California is
good, problem areas do exist. Point sources are seen as
under control, both in terms of number and severity, as a
result of construction and operation of waste treatment
facilities. Nonpoint sources are viewed as widespread,
difficult to define, and related to established land use
practices. Logging has contributed debris and sediments to
surface waters. Increasing mineralization of groundwater and
of surface water (Colorado River water used by the Imperial
and Coachella Valley agriculture), pesticide residues, and

heavy metals concentrations are major problems (Geraghty et al. 1973; U.S. Environmental Protection Agency 1977b). Elevated levels of mercury have been reported in the San Francisco Bay area and in the Merced River (Geraghty et al. 1973). Saltwater intrusion has degraded the quality of formerly usable aquifers along the southern coast (U.S. Environmental Protection Agency 1977b).

More than 139,000 km^2 (53,670 sq miles), or 45%, of the land area within the California region is underlain by groundwater reservoirs, both undeveloped and developed. Undeveloped reservoirs are those from which total annual withdrawal is less than 1.0×10^7 m^3 (8100 acre-ft). Groundwater reservoirs are listed for each of the nine subregions in Tables XXI and XXII). Usable reservoir volumes have been estimated for 37 of the 52 recognized groundwater reservoirs (Table XXIII); total usable reservoir capacity is at least 2.1×10^{10} m^3 (17×10^6 acre-ft). Reservoirs are of two main types: alluvial and volcanic bedrock. The aquifers are the most reliable (and often the sole) source of potable water in many parts of the California region. The following data are from Thomas and Phoenix (1976).

Alluvial sediments in the various valleys and plains comprise the most abundant groundwater reservoirs. Basaltic volcanic rocks, however, are prolific aquifers in the Cascade Range and Modoc Plateau areas, which encompass approximately 15% of the region. Consolidated rocks and residuum associated with the Sierra Nevada, Coast Ranges, and Basin Ranges are only locally important as groundwater reservoirs.

The alluvial aquifers in the Central Valley (Sacramento and San Joaquin basins subregions) occupy 10% of the region's land area. The predominantly sand and gravel water-bearing units are separated by silt and clay deposits, locally forming confined aquifers. Depths to water and pumpages from individual aquifers are given in Table XXII.

Folded and faulted sedimentary and metamorphic rocks constitute the Coast Ranges, which can be divided into four subregions: North Coastal, Central Coastal, South Coastal, and San Francisco Bay (Tables XXI and XXII). The aquifers consist of stream and valley alluvium and faulted consolidated rocks.

The North and South Lahontan and Colorado Desert subregions are the most arid in the region, including the Mojave and Colorado deserts and the dry valleys of the Great Basin. The groundwater reservoirs are comprised of the valley-floor and alluvial fan deposits.

Thick sequences of lava flows and tuffs and volcanic cones occur in the northeastern portion of the region. Especially in the valley areas, such as the Klamath and Sacramento river basins, these volcanics are permeable enough to be considered prolific aquifers. The volcanic rocks

Table XXI. Summary of the groundwater reservoirs

	Undeveloped groundwater reservoirs		Developed groundwater reservoirs			
Subregion	Number	Estimated area [km² (sq mi)]	Number	Area [km² (sq mi)]	Usable capacity [km³ (10⁶ acre-ft)]	Annual pumpage [km³ (10⁶ acre-ft)]
North Coastal	6	2,300 (888)	7	2,800 (1,080)		0.2 (0.2)
San Francisco Bay	1	100 (39)	9	4,000 (1,540)	3 (2.4)	0.4 (0.3)
Central Coastal	1	700 (270)	10	7,900 (3,050)	9 (7.3)	1.0 (0.8)
South Coastal	0		13	7,800 (3,010)	26 (21.1)	1.7 (1.4)
Coastal basins	8	3,100 (1,200)	39	22,500 (8,690)	38 (30.8)	3.3 (2.7)
Tributary valleys	7	2,300 (888)	2	1,400 (541)		
Sacramento Basin				10,800 (4,170)		3.1 (2.5)
Delta Area			1	3,100 (1,200)		1.2 (1.0)
San Joaquin Basin				13,000 (5,020)		8.0 (6.5)

Table XXI (continued)

Subregion	Undeveloped groundwater reservoirs		Developed groundwater reservoirs			
	Number	Estimated area [km² (sq mi)]	Number	Area [km² (sq mi)]	Usable capacity [km³ (10⁶ acre-ft)]	Annual pumpage [km³ (10⁶ acre-ft)]
Tulare Basin				11,300 (4,360)		3.7 (3.0)
Central Valley	7	2,300 (888)	3	39,600 (15,300)	125 (101)	16 (13)
North Lahontan	3	3,400 (1,310)	0			
South Lahontan	38	18,800 (7,260)	10	14,100 (5,440)	43 (34.9)	0.7 (0.6)
Colorado Desert	23	22,200 (8,570)	3	2,200 (849)	4 (3.2)	0.2 (0.2)
Interior basins	64	44,400 (17,100)	13	16,300 (6,290)	47 (38.1)	0.9 (0.7)
Total	79	50,000 (19,300)	55	78,000 (30,100)	210 (170)	20 (16.2)

Source: Adapted from Thomas and Phoenix (1976).

Table XXII. Developed groundwater reservoirs

County	Groundwater reservoir	Aquifer			Withdrawals from wells		
		Area [km² (sq mi)]	Depth zone [m (ft)]	Usable capacity [km³ (10³ acre-ft)]	Year	Pumpage [km³ (10³ acre-ft)]	Range in dissolved solids (mg/liter)[a]
	North Coastal subregion						
Del Norte, CA	Smith River Basin:						
	Smith River plain	180 (69.5)	3-11 (10-36)	0.09 (70)	1968	7 (5.68)	30-200
Klamath, OR	Klamath River Basin:						
	Sprague River valley	440 (170)			1970	31 (25.10)	80-230
	Swan Lake Valley	120 (46.3)			1970	31 (25.10)	80-270
	Yonna Valley	120 (46.3)			1970	16 (13.0)	110-270
	Lower Klamath River valley	490 (189)			1970	100 (81.10)	130-880
	Lower Klamath River valley	410 (158)			1954	12 (9.73)	80-830
Siskiyou, CA	Closed basin:						
	Butte Valley	470 (181)			1953	26 (21.10)	110-1,900
Humboldt, CA	Eel River and Mad River basins:						
	Eureka plain	600 (232)	3-12 (10-39)	0.15 (120)	1962	18 (14.6)	50-2,000
	Mad River valley (1-8),						
	Eel River valley (1-10)						
	San Francisco Bay subregion						
Mendocino, CA	Russian River basin:						
	Ukiah Valley	180 (69.5)	3-15 (10-50)	0.04 (30)	1954	12 (9.73)	110-1,120
Sonoma, CA	Santa Rosa Valley and Healdsburg area	470 (181)	3-60 (10-200)	1.2 (970)	1954	22 (17.80)	90-800
	San Francisco Bay:						
	Petaluma Valley	340 (131)	3-60 (10-200)	0.25 (200)	1958	2 (1.62)	110-4,800
Napa, CA	Napa Valley	210 (81.1)	3-60 (10-200)	0.30 (240)	1950	7 (5.68)	100-5,000
Sonoma, CA	Sonoma Valley	100 (38.6)	5-60 (16-200)	0.05 (40)	1950	2 (1.62)	130-2,800
Solano, CA	Suisan-Fairfield Valley	670 (259)	3-60 (16-200)	0.05 (40)	1949	10 (8.11)	300-1,350
Contra Costa, CA	Pittsburg plain	100 (38.6)	30-60 (100-200)	0.15 (120)	1931	10 (8.11)	480-2,060
	Clay Valley	160 (61.8)	6-60 (20-200)		1930	10 (8.11)	210-2,170
	Ygnacio Valley (2-6)						

Table XXII (continued)

County	Groundwater reservoir	Area [km² (sq mi)]	Aquifer		Withdrawals from wells		
			Depth zone [m (ft)]	Usable capacity [km³ (10³ acre-ft)]	Year	Pumpage [km³ (10³ acre-ft)]	Range in dissolved solids (mg/liter)[a]
	San Francisco Bay subregion						
Santa Clara, CA	Santa Clara Valley:						
	South Bay	830 (320)	8-60 (26-200)	0.95 (770)	1969	220 (178)	240-960
	East Bay	470 (181)			1970	50 (40.60)	300-7,000
Alameda, CA	Livermore Valley: Sunol Valley (2-11), San Ramon Valley (2-7)	570 (220)	8-60 (26-200)	0.25 (200)	1970	25 (20.30)	290-2,800
	Central Coastal subregion						
Santa Cruz, CA	Soquel Creek basin: Soquel-Aptos area	260 (100)					180-700
Santa Clara, CA	Pajaro River basin: Pajaro Valley	360 (139)	6-90 (20-300)	0.03 (20)	1969	65 (52.70)	170-1,500
	Llagas Valley	210 (81.1)			1969	60 (48.70)	250-550
San Benito, CA	Hollister Valley	670 (259)	6-60 (20-200)	1.0 (810)	1960	135 (109)	280-2,550
Monterey, CA	Salinas River basin: Salinas Valley	1,810 (699)	6-60 (20-200)	1.6 (1300)	1960	370 (300)	240-3,000
San Luis Obispo, CA	Paso Robles	2,330 (900)	15-75 (50-250)	2.1 (1700)	1967	55 (44.6)	
	Santa Maria River basin: Arroyo Grande Valley	100 (38.6)	30-240 (100-790)	0.15 (120)	1967	20 (16.2)	200-2,900
Santa Barbara, CA	Santa Maria Valley	520 (201)	6-60 (20-200)	1.2 (970)	1967	140 (114)	230-3,200
	Cuyama Valley	600 (232)		0.5 (400)	1967	80 (64.9)	400-5,000
	San Antonio Creek basin: San Antonio Creek valley	230 (88.8)		0.35 (280)	1967	14 (11.4)	300-3,000
Santa Barbara, CA	Santa Ynez River basin: Santa Ynez River valley	670 (259)	6-75 (20-250)	1.2 (970)	1967	50 (40.6)	400-2,000
	Santa Barbara Coastal basins: Santa Barbara basin Goleta Basin (3-16) Carpinteria Basin (3-18)	100 (38.6)	15-75 (50-250)	0.2 (160)	1967	9 (7.3)	340-1,400

Table XXII (continued)

South Coastal subregion

County	Groundwater reservoir	Area [km² (sq mi)]	Depth zone [m (ft)]	Usable capacity [km³ (10³ acre-ft)]	Year	Pumpage [km³ (10³ acre-ft)]	Range in dissolved solids (mg/liter)[a]
Ventura, CA	Santa Clara River Valley	1,190 (459)		0.75 (610)	1951		270-4,700
Los Angeles, CA	Los Angeles River and Santa Ana River Basins:						
	Coastal plain	1,300 (502)	WT-370 (WT-1210)	4.9 (3,970)	1970	350 (284)	140-1,340
	Central Basin	(600) (232)			1970	275 (223)	
	West Basin	(410) (158)			1970	75 (608)	
	San Fernando Valley	520 (201)	0-490 (0-1610)	11 (8,920)	1970	135 (109)	220-2,130
	San Gabriel Valley	520 (201)	6-460 (20-1510)	1.2 (970)	1965	250 (203)	110-1,000
	Raymond Basin	(100) (38.6)			1961-69	35 (28.4)	150-700
San Bernardino, CA	Upper Santa Ana Valley	1,680 (649)			1965	630 (511)	100-1,000
	Bunker Hill-San Timoteo	540 (208)		2.1 (1,700)			
	Chino-Riverside	1,110 (429)	6-210 (20-690)	6.8 (5,520)			
	Coastal Plain	930 (359)			1970	240 (195)	200-2,000
Orange, CA	San Jacinto River basin: San Jacinto basin	650 (251)					280-3,900
San Diego, CA	Santa Margarita and adjacent basins:						
	Lower Valley	60 (23.2)	2-msl (7-msl)	0.07 (60)	1966	11 (8.92)	180-1,600
	San Mateo (9-2)						
	San Onofre (9-3)						
	Temecula Valley	100 (38.6)		0.65 (530)	1961	12 (9.73)	250-5,000
	Warner Valley (9-8)						
	San Luis Rey basin:						
	San Luis Rey Valley	100 (38.6)	6-35 (20-115)	0.06 (50)			300-9,000
	San Dieguito River basin:						
	Escondido area (9-9)	210 (81.1)					250-5,000
	San Pasqual Valley (9-10)						
	San Diego River basin:						
	San Diego area	130 (50.2)	3-35 (10-115)	0.12 (100)			160-4,500
	El Cajon (6-16)						
	Warner Valley	100 (38.6)	6-35 (20-213)	0.07 (60)			150-420

Table XXII (continued)

County	Groundwater reservoir	Area [km^2 (sq mi)]	Depth zone [m (ft)]	Usable capacity [km^3 (10^3 acre-ft)]	Year	Pumpage [km^3 (10^3 acre-ft)]	Range in dissolved solids (mg/liter)[a]
	Central Valley subregion						
Lake, CA	Sacramento River basin: Kelseyville Valley, Upper Lake (5-13), Scott Valley (5-14), Burns Valley (5-17)	130 (50.2)	3-30 (10-100)	0.09 (70)	1951	17 (13.8)	80-660
Shasta, CA	Redding basin	1,300 (502)	6-60 (20-200)	0.15 (120)	1955		120-1,700
Several, CA	Sacramento Valley Mokelumne area:	11,000 (4,250)	6-60 (20-200)	35 (28,400)	1964	3,080 (2,500)	110-2,800
San Joaquin, CA	Delta	3,110 (1,200)	6-60 (20-200)		1966	1,230 (998)	300-3,500
Several, CA	San Joaquin River basin: San Joaquin Valley	13,000 (5,020)	6-60 (20-200)	69 (56,000)	1966	8,010 (6,500)	90-5,000
Kern, Kings, and Tulare, CA	Tulare close basin: Tulare Basin	11,700 (4,520)	6-60 (20-200)	46 (37,300)	1966	3,700 (3,000)	120-2,400
	South Lahontan subregion						
Inyo and Mono, CA	Closed basins: Owens Valley	2,230 (861)			1970	40 (32.4)	100-brine
San Bernardino, CA	Lower Mojave River valley	780 (301)	0-90 (0-300)	5.8 (4,700)	1963	85 (68.9)	190-2,340
	Middle Mojave River valley	1,090 (421)	0-90 (0-300)	11 (8,920)	1963	75 (60.8)	140-3,900
	Upper Mojave River valley	1,550 (598)	0-90 (0-300)	9.9 (8,030)	1963	55 (44.6)	80-2,760
Kern and Los Angeles, CA	Antelope Valley Divided: Cummings Valley:	4,140 (1,600)	6-60 (20-200)	6.7 (5,430)	1960	380 (308)	120-7,700
Kern, CA	Tehachapi Valley West, Tehachapi Valley East	130 (50.2)			1961	20 (16.2)	350-570
San Bernardino, CA	Closed basins: Fremont Valley	850 (328)	6-65 (20-213)	8.6 (6,970)	1958	40 (32.4)	350-brine
	Harper Valley	1,320 (510)			1963	15 (12.2)	320-10,700
Inyo and San Bernardino, CA	Searles Valley	650 (251)			1962	6 (4.87)	8,000
		100 (38.6)			1962	12 (9.73)	35,000
Inyo, Kern, and San Bernardino, CA	Indian Wells Valley	1,350 (521)	6-65 (20-213)	0.89 (720)	1968	15 (12.2)	140-brine

Table XXII (continued)

County	Groundwater reservoir	Area [km² (sq mi)]		Aquifer		Withdrawals from wells		
				Depth zone [m (ft)]	Usable capacity [km³ (10³ acre-ft)]	Year	Pumpage [km³ (10³ acre-ft)]	Range in dissolved solids (mg/liter)[a]
			Colorado Desert subregion					
San Bernardino, CA	Closed basins: Lucerne Valley	830	(320)			1952	20 (16.2)	340–5,000
Riverside, CA	Coachella Valley: Upper valley	620	(239)	WT–20 (WT–65)	4.4 (3,570)	1958	55 (44.6)	149–1,000
	Artesian basin	520	(201)				130 (105)	750–3,200
San Diego, CA	Borrego Valley	260	(100)	3–60 (10–200)			12 (9.73)	290–1,480

[a]Includes the range in dissolved-solids concentration in observation wells of the California state-wide water quality monitoring program.

Source: Adapted from Thomas and Phoenix (1976).

Table XXIII. Undeveloped groundwater reservoirs [withdrawals less than 10^6 m³ (8100 acre-ft) per year]

County	Groundwater reservoir	Estimated area [km² (sq mi)]	Natural outflow[a]	Exploration date			Range in dissolved solids (mg/liter)
				Year	Number of wells	Depth explored [m (ft)]	
North Coastal subregion							
Klamath, OR[b]	Klamath River Basin:						
	Klamath Marsh	600 (232)	Williamson River	1970	12	205 (670)	50-110
	Langell Valley	180 (69.5)	Lost River	1970	10	150 (490)	130-170
	Upper Klamath Valley	750 (290)	Klamath River	1954	13	145 (480)	80-100
	Poe Valley	80 (30.9)	Lost River	1970	9	135 (440)	140-270
Siskiyou, CA	Shasta Valley	650 (251)	Shasta River	1953	50	215 (700)	160-980
	Scott River Valley	210 (81.1)	Scott River	1953	5	65 (210)	30-420
San Francisco Bay subregion							
Sonoma, CA	Russian River basin:						
	Alexander Valley	100 (38.6)	Russian River	1954	40	135 (440)	220-1,300
Central Coastal subregion							
San Luis Obispo, CA	Closed basin:						
	Carrizo Plain	700 (270)	ET-Soda Lake	1954	8	305 (1000)	340-4,300
Sacramento Basin subregion							
Lake, OR[b]	Goose Lake Valley	410 (158)	ET-Goose Lake	1948	10	915 (3000)	120-1,460
Modoc, CA[b]		490 (189)		1964	15	230 (750)	100-450
Modoc, CA[b]	Pit River basin:						
	Alturas basin	230 (88.8)	Pit River	1964	25	305 (1000)	150-500
Modoc and Shasta, CA	Big Valley	260 (100)		1964	25	365 (1200)	150-1,380
Shasta, CA	Fall River Valley	260 (100)	Fall River	1964	50	215 (700)	100-550
Plumas and Sierra, CA	Feather River basin:						
	Sierra Valley	360 (139)	Feather River	1964	20	425 (1390)	120-1,400

Table XXIII (continued)

County	Groundwater reservoir	Estimated area [km² (sq mi)]	Natural outflow[a]	Year	Number of wells	Depth explored [m (ft)]	Range in dissolved solids (mg/liter)
						Exploration date	
San Joaquín Basin subregion							
San Benito, CA	Panoche Valley	130 (50.2)	Panoche Creek			120 (390)	
Kern, CA	Kern River Valley	70 (27.0)	Kern River	(partly overlain by Isabella Reservoir)			
North Lahontan subregion							
Modoc, CA[b]	Closed basins: Surpise Valley	910 (351)	ET-Alkali Lakes	1954	60	245 (800)	165-2,000
Washoe, NV[b]		30 (11.6)					
Modoc, CA	Madeline Plains	700 (270)	ET-Tule Lake	1964	10	260 (850)	100-245
Lassen, CA[b]	Honey Lake Valley	1,270 (490)	ET-Honey Lake	1964	100	425 (1,390)	175-1,350
Washoe, NV[b]		490 (189)		1967	10	120 (390)	170-5,000
South Lahontan subregion							
Mono, CA[b]	Closed basins: Mono Valley	520 (201)	ET-Mono Lake	1960	2	290 (950)	2,000-brine
Mineral, NV[b]		80 (30.9)					
Mono, CA	Adobe Lake Valley	160 (61.8)	ET-Adobe Lake	1962	4	10 (30)	130-280
	Long Valley	260 (100)	Lake Crowley	1954	8	25 (80)	90-1,500
Inyo, CA	Centennial (Black Springs) Valley	130 (50.2)	GW	1954			360
	Deep Springs Valley	100 (38.6)	ET-Deep Springs Lake	1955	4	235 (770)	
	Eureka Valley	410 (158)	ET-playa	1955	1	115 (380)	550
	Saline Valley	540 (208)	ET-Salt Lake	1955	1	15 (50)	3,700-brine
Inyo and San Bernardino, CA	Death Valley	3,420 (1,320)	ET-Badwater	1961	7	305 (1,000)	550-brine
San Bernardino, CA	Wingate Valley	180 (69.5)		1953			660
Inyo, CA[b]	Middle Amargosa basin	1,350 (521)	ET-playa	1962	20	145 (480)	300-2,900
San Bernardino, CA	Lower Kingston (Valjean) Valley	750 (290)	GW	1954	1	No water to 430 (130)	5,300-8,600

97

Table XXIII (continued)

County	Groundwater reservoir	Estimated area [km² (sq mi)]		Natural outflow[a]	Exploration date			Range in dissolved solids (mg/liter)
					Year	Number of wells	Depth explored [m (ft)]	
				South Lahontan subregion				
San Bernardino, CA	Upper Kingston (Shadow) Valley	700	(270)	GW	1961	10	120 (390)	340-1,1000
	Riggs Valley	260	(100)	GW	1954	1	45 (150)	1,740
	Red Pass Valley	390	(151)	ET-Red Pass Lake	1944	1	115 (380)	
	Bicycle Valley	310	(120)	ET-Bicycle Lake	1955	1	135 (440)	610-brine
	Avawatz Valley	180	(69.5)	ET-Drinkwater Lake				300-700
	Leach Valley	180	(69.5)	ET-Leach Lake				
Inyo, San Bernardino, CA	Mesquite Valley	310	(120)	ET-Mesquite Lake	1959	22	335 (1,100)	300-6,300
San Bernardino, CA[b]	Ivanpah Valley	780	(301)	ET-Ivanpah Lake	1960	19	360 (1,180)	290-2,200
San Bernardino, CA	Kelso Valley	960	(371)	GW	1961	2	195 (640)	250-750
	Broadwell Valley	310	(120)	ET-Broadwell Lake	1883	1	330 (1,080)	470-1,260
	Soda Lake Valley	1,530	(591)	ET-Soda Lake GW	1961	20	150 (490)	240-3,400
	Silver Lake Valley	100	(38.6)	ET-Silver Lake GW	1954	3	55 (180)	1,100-1,740
	Cronise Valley	390	(151)	ET-Cronise Lake	1961	13	230 (750)	450-3,100
	Langford Valley	130	(50.2)	ET-Langford Lake	1958	7	160 (520)	470-640
	Coyote Lake Valley	390	(151)	ET-Coyote Lake	1961	5	175 (570)	300-2,500
	Caves Canyon Valley	260	(100)	GW	1961	5	65 (210)	200-1,300
	Troy Valley	340	(131)	ET-Troy Lake	1961	20	120 (390)	280-6,500
	El Mirage Valley	310	(120)	GW	1961	75	295 (970)	320-2,600
	Superior Valley	440	(170)	ET-Superior Lake	1956	25	115 (380)	360-2,300
	Cuddeback Valley	340	(131)	ET-Cuddeback Lake	1956	30	90 (300)	370-3,700
	Pilot Knob Valley	520	(201)	GW	1958	1		400
Inyo, CA	Coso Valley	130	(50.2)	GW	1946	1	35 (110)	150-1,300
	Rose Valley	160	(61.8)	ET-playa		6	55 (180)	350-750
	Darwin Valley	180	(69.5)		1954	3	75 (250)	780-brine
	Panamint Valley	930	(359)	ET-Panamint Lake	1955	3	300 (980)	
	Brown Mt. Valley (6-76)							
San Bernardino, CA	Lost Lake Valley	100	(38.6)	ET-Lost Lake	1953	0		360
Inyo and San Bernardino, CA	California Valley	210	(81.1)	ET-playa	1953	2	15 (50)	350-500

Table XXIII (continued)

County	Groundwater reservoir	Estimated area [km² (sq mi)]		Natural outflow[a]	Exploration date		Depth explored [m (ft)]		Range in dissolved solids (mg/liter)
					Year	Number of wells			
	Colorado Desert subregion								
San Bernardino, CA	Closed Basins:								
	Lanfair Valley	730	(282)	GW	1952	18	270	(890)	230–2,000
	Fenner Valley	1,860	(718)	GW	1952	5	285	(940)	280–870
	Ward Valley	1,990	(768)	ET–Danby Lake	1952	2	315	(1,030)	330–brine
Riverside, CA	Chuckwalla Valley	2,250	(869)	ET–Palm Lake	1952	10	185	(610)	270–brine
	Pinto Basin	800	(309)	GW	1937	3	170	(560)	120–830
San Bernardino, CA	Cadiz Valley	1,110	(429)	ET–Cadiz Lake	1910	1	105	(340)	610–brine
	Bristol Valley	1,840	(710)	ET–Bristol Lake	1910	8	695	(2,280)	290–brine
	Dale Valley	670	(259)	ET–Dale Lake	1952	10			1,060–brine
	Twentynine Palms Valley	960	(371)	GW	1952	50	150	(490)	100–1,180
	Copper Mt. Valley (7–11)								
	Warren Valley (7–12)								
	Deadman Valley	750	(290)	GW	1952	35			180–600
	Lavic Valley	100	(38.6)	ET–Lavic Lake	1917	1	40	(130)	1,680
	Ames Valley (7–16)								
	Bessemer Valley	180	(69.5)	ET–Galway Lake					
	Means Valley	100	(38.6)	ET–Means Lake					
	Johnson Valley	360	(139)	GW	1952	1	45	(150)	340–610
Imperial, CA	West Salton Sea basin	750	(290)	Salton Sea	1950	3			2,260–brine
	Clark Valley	100	(386)	ET–Clark Lake					
San Diego, CA	Ocotillo Valley	210	(81.1)	GW	1952	6	65	(210)	700
	San Felipe (Earthquake) Valley	160	(61.8)	San Felipe Creek	1952	3			1,060
	Vallecito and Carrizo Valley	310	(120)	GW					
Imperial, CA	Coyote Wells Valley	260	(100)	GW	1948	6	50	(160)	440–8,700
	Imperial Valley	4,840	(1,870)	Salton Sea		80	335	(1,100)	690–7,500
Riverside, CA	Orocopia Valley	360	(139)	GW	1952				350–1,500
	Chocolate Valley	310	(120)	GW					350–brine
Riverside and Imperial, CA	East Salton Sea basin	1,170	(452)	Salton Sea	1952	4	100	(330)	350–3,850

[a] Natural outflow: ET, evapotranspiration; GW, outflow to groundwater reservoirs.

[b] Interstate reservoir; data given apply only to part of reservoir in that state.

Source: Adapted from Thomas and Phoenix (1976).

99

constitute an area of more than 26,000 km^2 (10,000 sq miles) in the North Coastal, Sacramento River, and North Lahontan subregions (Tables XXI and XXII).

The quality of groundwater generally decreases (as does total precipitation) from north to south within the region. Within any one subregion, the quality varies over a wide range from aquifer to aquifer (Tables XXI and XXII). The TDS concentrations in groundwaters of the North Coastal subregion vary from 30 to 2000 mg/liter. The Central Coastal subregion has TDS concentrations between 170 and 5000 mg/liter. Other subregional areas and ranges in groundwater TDS concentrations (in mg/liter) are: San Francisco Bay, 90 to 7000; South Coastal, 100 to 9000; Central Valley (Sacramento and San Joaquin), 80 to 5000; North Lahontan, 100 to 5000; South Lahontan, 80 to 350,000; Colorado Desert, 100 to 35,000.

Recharge to the groundwater reservoirs takes place through infiltration of precipitation in the form of rainfall and of snowmelt from the high mountains. Some recharge occurs from downward-percolating irrigation water. Discharge takes place primarily through springflow, evapotranspiration, and underflow to lower valleys.

Groundwater in the California region is used most extensively for irrigation. Additional uses are for domestic, municipal, and industrial uses (Thomas and Phoenix 1976).

In the California Water Resource Region, average annual runoff is equivalent to 230 × 10^6 m^3/day (62 Bgd) (U.S. Geological Survey 1977). Precipitation is subject to large seasonal, year-to-year, and geographic variation (Water Resources Council 1974). Of the total off-channel water withdrawals of 190 × 10^6 m^3/day (51 Bgd) in 1975, surface freshwater withdrawals accounted for about 83 × 10^6 m^3/day (22 Bgd), fresh groundwater about 72 × 10^6 m^3/day (19 Bgd), and saline surface water about 38 × 10^6 m^3/day (10 Bgd). About 56% of freshwater withdrawn was consumed. Irrigation is the greatest water user and consumer of freshwater in the region (Table I) (U.S. Geological Survey 1977). Colorado River water via the All-American and Coachella canals serves about 240,000 irrigable hectares (600,000 acres) in the Salton Sea basin (Imperial Irrigation District). Thermoelectric power (condenser and reactor cooling) accounts for most of the saline surface-water use, whereas irrigation accounts for most of the groundwater use (U.S. Geological Survey 1977). Hydroelectric power generation uses about 280 × 10^6 m^3/day (74 Bgd), with the California region ranking ahead of seven other regions in the conterminous United States in this respect (U.S. Geological Survey 1977).

Temporal and spatial variation in water availability has been largely overcome by water development projects, especially the California State Water Project and Central Valley Project. Eventual increased competition for water is

forecast, with possible limitations on agricultural development. However, the extensive water development of the
region, coupled with increased use of lower quality waters for
energy technologies, is seen as means of providing generally
adequate water supplies for the region (Water Resources
Council 1974).

Aquatic Ecology

    The discussion of biotic resources of the Upper Colorado
WRR is applicable to the California WRR.

PACIFIC NORTHWEST

Hydrology, Water Quality, and Water Use

    The Pacific Northwest Water Resource Region, with an
area of about 700,000 km^2 (271,000 sq miles) (U.S. Geological
Survey 1977), includes the Columbia River and its major
tributaries (Snake, Owyhee, Salmon, Deschutes, Willamette,
Clark Fork, Kootenai, and Pend Oreille rivers), as well as
other coastal rivers in Washington and Oregon (Skagit,
Chehalis, and Rogue rivers). Large natural lakes include
Flathead, Coeur d'Alene, Pend Oreille, and Jackson lakes in
the headwaters of the Columbia River system; other natural
lakes such as Malheur and Harney lakes are in the part of
southeastern Oregon that is physiographically in the Great
Basin. Crater Lake formed in the collapsed volcanic cone of
Mount Mazama. The Columbia River system includes major
impoundments, such as Franklin Roosevelt Lake (one of the ten
largest in the nation in terms of storage capacity) (Geraghty
et al. 1973) on the Columbia River and other major reservoirs
on the tributaries. The Skagit River also includes a major
reservoir, Ross Lake. Coastal areas include the mouth of
the Columbia River, Puget Sound, and numerous other bays in
Washington and Oregon.
    Surface waters of the Pacific Northwest WRR are typically
soft (<60 mg/liter hardness as CaCO$_3$) in Washington, Oregon,
and northern Idaho. Moderate hardness (60 to 120 mg/liter)
is more common along the Columbia River above the mouth of
the Snake River, in the Snake-Owyhee boundary area between
Oregon and Idaho, and in most of southern Idaho. Hard waters
(120 mg/liter and above) are found in southeastern Idaho and
in the Snake River near Boise, Idaho. Levels of TDS are
generally lowest (<120 mg/liter) in Washington, western
Oregon, and northern Idaho. Moderate levels (120 to 350 mg/
liter) are found in southern Idaho, eastern Oregon, and areas
in eastern Washington. Concentrations exceeding 350 mg/liter

are found in isolated areas in southeastern Oregon (physio-graphically in the Great Basin) and central Washington. Levels of TSS are generally low (<270 mg/liter) in Washington, western and southern Oregon, northern Idaho, and some impounded and headwater reaches of the upper Snake River system. Highest concentrations (>1900 mg/liter) are found in parts of the Snake River Plain in Idaho. Intermediate levels (270 to 1900 mg/liter) are found along the Snake River main-stem and other parts of southern Idaho and in areas of eastern and central Oregon (Geraghty et al. 1973).

Nonpoint-source discharges are important in regional water quality degradation. Mineralization from mineral mining (Idaho), accelerated runoff from silviculture, and agricultural runoff (including pesticide residues) are major problems. Dissolved oxygen (DO) levels in the Willamette River, formerly stressed by ammonia wastes, are now under control. Several lakes in eastern Washington are eutrophic; in many cases, the lakes are naturally in advanced succes-sional stages, although nonpoint-source pollution is also implicated. The Boise River and Middle Chehalis River are cited by the EPA as cases in which waste treatment has resulted in water quality improvement (U.S. Environmental Protection Agency 1977b).

The Pacific Northwest region has groundwater supplies obtainable from igneous intrusive, igneous extrusive, meta-morphic, and consolidated sedimentary rocks and from uncon-solidated sediments. In general, well yields are small; however, some large yields have been reported.

The aquifers in the mountain range area are predominantly crystalline, metamorphic, and sedimentary rocks. The producing zones are coincident with fractures, joints, and faults. Residual material also produces small well yields. The best well yields are obtained from sand and gravel sediments in the intermontane valleys (Walton 1970).

Much of the region is underlain by lava flows, which are interbedded with and mantled in places by fluvial and lacustrine sediments. Individual flows range in thickness from less than 10 m to more than 100 m. The entire volcanic sequence varies between tens of meters to more than several hundred meters thick. The best production zones in the basalts are those in which lava tubes, abundant vesicles, shrinkage cracks, and unconformities exist. Permeable sands and gravels in the section also yield water to wells. The highest yielding aquifers are unconsolidated deposits along stream valleys, some of which are of glacial outwash origin (Walton 1970).

In the Pacific Northwest Water Resource Region, average annual runoff is equivalent to 800×10^6 m^3/day (210 Bgd). Of the total off-channel water withdrawal of 130×10^6 m^3/day (33 Bgd) in 1975, surface freshwater withdrawals accounted for 99×10^6 m^3/day (26 Bgd); fresh groundwater contributed

the remainder. About 33% of freshwater withdrawn was consumed (U.S. Geological Survey 1977). Irrigation accounts for
the major portion of water use and freshwater consumption
(Table I). Irrigation and, to a lesser extent, industry
dominate groundwater use. The Pacific Northwest region uses
5700×10^6 m^3/day (1500 Bgd) for hydroelectric power, almost
half the national total, and the greatest use of any region
in the nation (U.S. Geological Survey 1977). Because of
freezing conditions, runoff in the Columbia River system is
lowest in the winter when power loads are highest; streamflow
regulation has largely compensated for this imbalance (Water
Resources Council 1974). The Water Resources Council has
projected a general availability of energy-related water
supplies in the foreseeable future, particularly for the
western portion of the region where major load centers are
located (Water Resources Council 1974).

Aquatic Ecology

This WRR contains many of the same habitats described
for the Texas Gulf WRR. Much of the area is mountainous, and
much of it is arid. The greatest physiographic difference
between this area and the preceding section (California) is the
occurrence of lower average temperatures.

The upland areas in the Pacific Northwest contain
mountain lakes and streams similar to those described in
previous sections. Likewise, saline lakes and playas much
like those in adjacent water resource regions occur in the
more arid areas (Edmonson 1966).

Before being extensively impounded, the principal rivers
in the region (e.g., the Snake and Columbia rivers) had
characteristics similar to those of the upper Colorado
River — high current velocities and greatly fluctuating
discharge volumes. In their original state, these rivers
contained few phytoplankton or periphyton, had limited
populations of benthic invertebrates (except in the headwaters), and contained large seasonal populations of salmon.
Impounding has greatly stabilized flows and has significantly
altered the biota. Dense plankton populations now develop in
certain portions of the rivers, and the bottom fauna is rich
and dense (standing crops greater than 2500 organisms per
square meter have been reported). A fishery has developed
that is more typical of warm, static waters than cool
riverine areas, and the salmon spawning runs have been
greatly impeded by the dams (Edmonson 1966).

In the coastal areas precipitation is considerably
greater than in the interior regions and a few soft, oligotrophic lakes occur (Edmonson 1966). Estuaries in this WRR
(and along the California coast) serve as important spawning
and nursery areas for many marine organisms. However,

because of low temperatures and fewer coastal marshes, Pacific coast estuaries are much less productive than those emptying into the Gulf of Mexico (Reid 1961).

Major anthropogenic changes that have altered the biota in this WRR include channelization and impounding of many of the rivers; the addition of agricultural runoff that has contributed pesticides, nutrients, and salts; and municipal waste discharges that have exacerbated eutrophication (Edmonson 1966). Some of the areas that have been severely affected have been restored at least partially. Lake Washington, for example, has experienced a slowing in the rate of eutrophication by extensive lake renewal and waste diversion efforts (Edmonson 1966). The trophic status of Lake Washington, as well as that of ultraoligotrophic Waldo Lake (Oregon) and mesotrophic Lake Sammamish (Washington), is described in U.S. Environmental Protection Agency (1977d).

3. ENDANGERED SPECIES

Table XXIV lists federally designated endangered and threatened aquatic flora and fauna in the conterminous United States. *Endangered* species are those in danger of extinction throughout all or a significant portion of their range. *Threatened* species are those that are likely to become endangered within the foreseeable future. Species designated as endangered or threatened are imperiled by any or all of the following: (1) destruction, modification, or curtailment of habitat or range; (2) overutilization for commercial, sporting, scientific, or educational purposes; (3) disease or predation; (4) regulatory inadequacies (lack of effective protection and conservation measures); or (5) other natural or manmade factors. As listed in Table XXIV, *critical habitats* have been designated for some of these species; such habitats are necessary for some or all stages of the species' life history and, consequently, for the survival and recovery of the species. In accordance with the Endangered Species Act of 1973 (PL 93-205), actions authorized, funded, or carried out by Federal agencies must not jeopardize the existence of endangered or threatened species or result in the destruction or modification of designated critical habitat. The capture, possession, transport, and sale of such species are also generally prohibited.

In addition to the species listed in Table XXIV, additional species are under review for possible protection under the Endangered Species Act; other species have been officially proposed as endangered or threatened pending final action. Changes in status have also been proposed. Some state-designated endangered species are not included on the Federal list.

Table XXIV. Federally designated endangered and threatened aquatic species
of the conterminous United States[a]

Species	Distribution	Water resource region
Fish		
Pahranagat bonytail, *Gila robusta jordani* (E)	NV	Great Basin
Alabama cavefish, *Speoplatyrhinus poulsoni* (T,CH)	AL	Tennessee
Humpback chub, *Gila cypha* (E)	AZ,UT,WY	Lower Colorado, Upper Colorado
Mohave chub, *Gila mohavensis* (E)	CA	California, Great Basin
Slender chub, *Hybopsis cahni* (T,CH)	TN,VA	Tennessee
Spotfin chub, *Hybopsis monacha* (T,CH)	VA,TN,NC	Tennessee
Longjaw cisco, *Coregonus alpenae* (E)	Lakes Michigan, Huron, and Erie	Great Lakes
Cui-ui, *Chasmistes cujus* (E)	NV	Great Basin
Kendall Warm Springs dace, *Rhinichthys osculus thermalis* (E)	WY	Upper Colorado
Moapa dace, *Moapa coriacea* (E)	NV	Lower Colorado
Bayou darter, *Etheostoma rubrum* (T)	MS	Lower Mississippi
Fountain darter, *Etheostoma fonticola* (E)	TX	Texas Gulf
Leopard darter, *Percina pantherina* (T,CH)	OK,AR	Arkansas–White–Red
Maryland darter, *Etheostoma sellare* (E)	MD	Mid–Atlantic
Okaloosa darter, *Etheostoma okaloosae* (E)	FL	South Atlantic–Gulf
Slackwater darter, *Etheostoma boschungi* (T,CH)	AL,TN	Tennessee
Snail darter, *Percina tanasi* (E,CH)	TN	Tennessee
Watercress darter, *Etheostoma nuchale* (E)	AL	South Atlantic–Gulf
Big Bend gambusia, *Gambusia gaigei* (E)	TX	Rio Grande
Clear Creek gambusia, *Gambusia heterochir* (E)	TX	Texas Gulf
Pecos gambusia, *Gambusia nobilis* (E)	TX	Rio Grande
Pahrump killifish, *Empetrichthys latos* (E)	NV	Great Basin
Scioto madtom, *Noturus trautmani* (E)	OH	Ohio
Yellowfin madtom, *Noturus flavipinnis* (T,CH)	TN,VA	Tennessee
Blue pike, *Stizostedion vitreum glaucum* (E)	Lakes Erie and Ontario	Great Lakes
Comanche Springs pupfish, *Cyprinodon elegans* (E)	TX	Rio Grande
Devil's Hole pupfish, *Cyprinodon diabolis* (E)	NV	Great Basin
Owens River pupfish, *Cyprinodon radiosus* (E)	CA	California, Great Basin
Tecopa pupfish, *Cyprinodon nevadensis calidae* (E)	CA	Great Basin
Warm Springs pupfish, *Cyprinodon nevadensis pectoralis* (E)	NV	Great Basin

Table XXIV (continued)

Species	Distribution	Water resource region
Fish (continued)		
Colorado River squawfish, *Ptychocheilus lucius* (E)	AZ,CA,CO,NM,UT,WY	Upper Colorado, Lower Colorado
Unarmored threespine stickleback, *Gasterosteus aculeatus williamsoni* (E)	CA	California
Shortnose sturgeon, *Acipenser brevirostrum* (E)	Atlantic coast	Mid-Atlantic, South Atlantic-Gulf
Gila topminnow, *Poeciliopsis occidentalis* (E)	AZ	Lower Colorado
Arizona trout, *Salmo apache* (T)	AZ	Lower Colorado
Gila trout, *Salmo gilae* (E)	NM	Lower Colorado
Greenback cutthroat trout, *Salmo clarki stomias* (T)	CO	Missouri Basin, Arkansas-White-Red
Lahontan cutthroat trout, *Salmo clarki henshawi* (T)	CA,NV	Great Basin
Little Kern golden trout, *Salmo aguabonita whitei* (T,CH)	CA	California
Paiute cutthroat trout, *Salmo clarki seleniris* (T)	CA	California, Great Basin
Woundfin, *Plagopterus argentissimus* (E)	AZ,NV,UT	Lower Colorado
Clams		
Alabama lamp pearly mussel, *Lampsilis virescens* (E)	AL	Tennessee
Appalachian monkeyface pearly mussel, *Quadrula sparsa* (E)	VA,TN	Tennessee
Birdwing pearly mussel, *Conradilla caelata* (E)	VA,TN	Tennessee
Cumberland bean pearly mussel, *Villosa trabilis* (E)	KY	Ohio
Cumberland monkeyface pearly mussel, *Quadrula intermedia* (E)	VA,TN	Tennessee
Curtis' pearly mussel, *Dysnomia florentina curtisi* (E)	MO	Arkansas-White-Red
Dromedary pearly mussel, *Dromus dromas* (E)	VA,TN	Tennessee
Fat pocketbook pearly mussel, *Potamilus capax* (E)	AR,MO	Arkansas-White-Red, Lower Mississippi
Finerayed pigtoe pearly mussel, *Fusconaia cuneolus* (E)	AL,VA,TN	Tennessee
Greenblossom pearly mussel, *Dysnomia torulosa gubernaculum* (E)	VA,TN	Tennessee
Higgins' eye pearly mussel, *Lampsilis higginsi* (E)	MN,WI,IL,MO	Upper Mississippi
Pale lilliput pearly mussel, *Toxolasma cylindrella* (E)	AL,TN	Tennessee
Pink mucket pearly mussel, *Lampsilis orbiculata orbiculata* (E)	AL,WV,TN,KY	Tennessee, Ohio

Table XXIV (continued)

Species	Distribution	Water resource region
Clams (continued)		
Rough pigtoe pearly mussel, *Pleurobema plenum* (E)	KY,VA,TN	Tennessee, Ohio
Sampson's pearly mussel, *Dysnomia sampsoni* (E)	IN,IL	Ohio
Shiny pigtoe pearly mussel, *Fusconaia edgariana* (E)	AL,VA,TN	Tennessee
Tan riffle-shell mussel, *Epioblasma walkeri* (E)	VA,TN,KY	Tennessee, Ohio
Tuberculed-blossom pearly mussel, *Dysnomia torulosa torulosa* (E)	KY,IL,TN,WV	Ohio, Tennessee
Turgid-blossom pearly mussel, *Dysnomia turgidula* (E)	TN	Tennessee
White cat's paw pearly mussel, *Dysnomia sulcata delicata* (including *D.s. perobliqua*) (E)	OH,MI,IN	Great Lakes
White warty-back pearly mussel, *Plethobasus cicatricosus* (E)	AL,TN	Tennessee
Yellow-blossom pearly mussel, *Dysnomia florentina florentina* (E)	TN	Tennessee
Orange-footed pimpleback, *Plethobasus cooperianus* (E)	AL,TN	Tennessee
Mammals		
Florida Manatee, *Trichechus manatus latirostris* (E,CH)	FL	South Atlantic—Gulf
Crustaceans		
Socorro Isopod, *Exophaeroma thermophilus* (E)	NM	Rio Grande

Sources: U.S. Department of the Interior (1971); U.S. Department of the Interior (1973); and U.S. Fish and Wildlife Service, *Fed. Regist.*, Sept. 25, 1975; June 14, 1976; July 14, 1977; Aug. 11, 1977; Aug. 23, 1977; Sept. 9, 1977; Sept. 22, 1977; Nov. 21, 1977; Jan. 27, 1978; Mar. 27, 1978; Apr. 13, 1978; Apr. 18, 1978; May 23, 1978; Jan. 17, 1979.

4. WILD AND SCENIC RIVERS

In accordance with the Wild and Scenic Rivers Act
(PL 90-542), selected river segments of the United States
have been designated as wild, scenic, and/or recreational.
These waters are to be preserved in a free-flowing condition,
and their values (scenic, recreational, geologic, fish and
wildlife, historical, or cultural) are to be protected. The
categorization as wild, scenic, or recreational is a function
of development and accessibility. *Wild* rivers are free of
impoundments, are generally inaccessible except by trail, and
have primitive watersheds or shorelines and unpolluted waters.
Scenic rivers are accessible, in places, by roads. *Recreational* rivers are readily accessible by road or railroad and
may have undergone some shoreline development as well as past
impoundment or diversion. In addition to the stream itself,
the designated area may include a corridor up to 0.8 km
(0.5 mile) in width (U.S. Department of the Interior 1977).
The river areas are to be administered so that the values
causing the areas to be included in the Wild and Scenic
Rivers System are protected. Primary emphasis is to be given
to protecting aesthetic, scenic, historical, archaeological,
and scientific features. Where appropriate, other uses that
do not interfere with these values will not be limited (Wild
and Scenic River Act 1976).

Table XXV lists the river segments currently included in
the National Wild and Scenic Rivers System; the state and
Water Resource Region (Fig. I) in which they are located; and
lengths of the wild, scenic, and recreational segments. In
addition to these 19 segments, others are under consideration
for inclusion or have been proposed for addition to the list.
Many states also have wild and scenic river programs.

Table XXV. Components of the National Wild and Scenic Rivers System

River	State	Water resource region	System (miles)[a]			
			Total	Wild	Scenic	Recreational
Clearwater, Middle Fork	ID	Pacific Northwest	185	54		131
Eleven Point	MO	Arkansas–White–Red	44		44	
Feather	CA	California	108	33	10	65
Flathead	MT	Pacific Northwest	219	98	41	80
Missouri	MT	Missouri Basin	159	72	28	59
Rapid	ID	Pacific Northwest	31	31		
Rio Grande	NM	Rio Grande	53	52		1
Rogue	OR	Pacific Northwest	85	33	8	44
Salmon, Middle Fork	ID	Pacific Northwest	104	103		1
Snake	ID,OR	Pacific Northwest	67	33	34	
Allagash	ME	New England	95	95		
Chattooga	NC,SC,GA	South Atlantic–Gulf	57	40	3	15
Little Beaver	OH	Ohio	33		33	
Little Miami	OH	Ohio	66		18	48
New	NC	Ohio	27		27	
Obed	TN	Tennessee	46	46		
St. Croix	MN,WI	Upper Mississippi	200		181	19
St. Croix, Lower	MN,WI	Upper Mississippi	52		12	40
Wolf	WI	Great Lakes	25		25	

[a] 1 mile = 1.6 km.

Source: U.S. Department of the Interior (1977).

Baker, E. T., Jr. and J. R. Wall. 1976. "Summary Appraisals of the Nation's Ground-water Resources — Texas-Gulf Region," U.S. Geological Survey Professional Paper 813-F.

Beck, K. C., J. H. Reuter and E. M. Perdue. 1974. "Organic and Inorganic Geochemistry of Some Coastal Plain Rivers of the Southeastern United States," *Geochim. Cosmochim. Acta* 38: 341-364.

Bedinger, M. S. and R. T. Sniegocki. 1976. "Summary Appraisal of the Nation's Ground-water Resources — Arkansas-White-Red Region," U.S. Geological Survey Professional Paper 813-H.

Beeton, A. M. and D. C. Chandler. 1966. "The St. Lawrence Great Lakes," in *Limnology in North America,* D. G. Frey, Ed., pp. 535-558. (Madison, WI: University of Wisconsin Press).

Berg, C. O. 1966. "Middle Atlantic States," in *Limnology in North America,* D. G. Frey, Ed., pp. 191-238. (Madison, WI: University of Wisconsin Press).

Bloyd, R. M., Jr. 1974. "Summary Appraisals of the Nation's Ground-water Resources — Ohio Region," U.S. Geological Survey Professional Paper 813-A.

Bloyd, R. M., Jr. 1975. "Summary Appraisals of the Nation's Ground-water Resources — Upper Mississippi Region," U.S. Geological Survey Professional Paper 813-B.

Brooks, J. L. and E. S. Deevey, Jr. 1966. "New England," in *Limnology in North America,* D. G. Frey, Ed., pp. 117-162. (Madison, WI: University of Wisconsin Press).

Carlander, K. D., R. S. Campbell and W. H. Irwin. 1966. "Mid-Continent States," in *Limnology in North America,* D. G. Frey, Ed., pp. 317-348. (Madison, WI: University of Wisconsin Press).

Cederstrom, D. J., E. H. Boswell and G. R. Tarver. 1979. "Summary Appraisals of the Nation's Ground-water Resources — South Atlantic Gulf Region," U.S. Geological Survey Professional Paper 813-0.

Christie, W. J. 1974. "Changes in the Fish Species Composition of the Great Lakes," *J. Fish. Res. Board Can.* 31: 827-854.

Christie, W. J., G. R. Spangler, D. A. Hurley and A. M. McCombie. 1972. Effects of Species Introductions on Salmonid Communities in Oligotrophic Lakes," *J. Fish Res. Board Can.* 29: 969-973.

Cole, G. A. 1966. "The American Southwest and Middle America," in *Limnology in North America*, D. G. Frey, Ed., pp. 393-434. (Madison, WI: Wisconsin Press).

DeBuchananne, G. D. and R. M. Richardson. 1956. "Ground-water Resources of East Tennessee," *Tennessee Div. Geology Bull.* 5B.

Dineen, C. F. 1953. "An Ecological Study of a Minnesota Pond," *Am. Midl. Nat.* 50: 349-376.

Dobson, J. E. and A. D. Shepherd. 1979. *Water Availability for Energy in 1985 and 1990.* ORNL/TM-6777, Oak Ridge National Laboratory, Oak Ridge, TN.

Eakin, T. E., D. Price and J. R. Harrill. 1976. "Summary Appraisals of the Nation's Ground-water Resources — Great Basin Region," U.S. Geological Survey Professional Paper 813-G.

Eddy, S. M. 1966. "Minnesota and the Dakotas," in *Limnology in North America*, D. G. Frey, Ed., pp. 301-317. (Madison, WI: University of Wisconsin Press).

Edmonson, W. T. 1966. "Pacific Coast and Great Basin," in *Limnology in North America*, D. G. Frey, Ed., pp. 371-392. (Madison, WI: University of Wisconsin Press).

Fassett, N. C. 1930. "The Plants of Some Northeastern Wisconsin Lakes," *Trans. Wis. Acad. Sci. Arts Lett.* 25: 155-168.

Fassett, N. C. 1957. *A Manual of Aquatic Plants.* (Madison, WI: University of Wisconsin Press).

Geraghty, J. J., D. W. Miller, F. van der Leeden and F. L. Troise. 1973. *Water Atlas of the United States*, 3rd ed. (Port Washington, N.Y.: Water Information Center, Inc.) 122 pp.

Gerking, S. D. 1950. "Populations and Expolitation of Fishes in a Marl Lake," *Invest. Indiana Lakes Streams* 2: 47-72.

Gerking, S. D. 1966. "Central States," in *Limnology in North America*, D. G. Frey, Ed., pp. 239-268. (Madison, WI: University of Wisconsin Press).

Goldman, C. R. 1972. "The Role of Minor Nutrients in Limiting the Productivity of Aquatic Ecosystems," in *Nutrients and Eutrophication*, American Society of Limnology and Oceanography, special symposia, Vol. 1, pp. 21-33.

Gumming, G. D. 1966. "Illinois," in *Limnology in North America*, D. G. Frey, Ed., pp. 163-190. (Madison, WI: University of Wisconsin Press).

Harris, B. B. and J. K. G. Silbey. 1940. "Limnological Investigation on Texas Reservoir Lakes," *Ecol. Monogr.* 10: 111-143.

Hornbeck, J. W., G. E. Likens and J. S. Eaton. 1977. "Seasonal Patterns in Acidity of Precipitation and their Implications for Forest Stream Ecosystems," *Water Air Soil Pollut.* 7: 355-365.

Hutchins, W. A. 1977. *Water Rights Laws in the Nineteen Western States*. Miscellaneous publication No. 1206. U.S. Department of Agriculture, Washington, D.C.

Hutchinson, G. E. 1967. *A Treatise on Limnology, Vol. II*. (New York: John Wiley and Sons, Inc.) 1115 pp.

Hynes, H. B. N. 1971. *The Biology of Polluted Waters*. (Toronto: University of Toronto Press) 202 pp.

Hynes, H. B. N. 1972. *The Ecology of Running Waters*. (Toronto: University of Toronto Press) 555 pp.

Imperial Irrigation District. 1977. "From Desert Wasteland to Agricultural Wonderland: The Story of Water and Power," Fact Booklet, El Centro, CA.

Imperial Irrigation District. "The Colorado River and Imperial Valley Soils," Bulletin No. 1075, El Centro, CA.

Johnson, W. E. and A. D. Hasler. 1954. "Rainbow Trout Production in Dystrophic Lakes," *J. Wildl. Manage.* 18: 113-134.

Jones, J. R., B. P. Borofka and R. W. Bachmann. 1976. "Factors Affecting Nutrient Loads in Some Iowa Streams," *Water Res.* 10: 117-122.

Juday, C. 1907. "Studies on Some Lakes in the Rocky and Sierra Nevada Mountains," *Trans. Wis. Acad. Sci. Arts Lett.* 15: 780-793.

Kilkus, S. P., J. D. La Perriere and R. W. Bachmann. 1975. "Nutrients and Algae in Some Iowa Streams," *J. Water Pollut. Control Fed.* 47: 1870-1979.

Klopatek, J. M., H. H. Shugart, Jr., W. F. Harris and R. W. Brocksen. 1979. *The Need for Regional Ecology.* ORNL/TM-6799, Oak Ridge National Laboratory, Oak Ridge, TN.

Knight, A., R. C. Ball and F. F. Hooper. 1962. "Some Estimates of Primary Production Rates in Michigan Ponds," *Pap. Mich. Acad. Sci. Arts Lett.* 47: 219-233.

Macan, T. T. 1974. *Freshwater Ecology,* 2nd ed. (New York: John Wiley and Sons) 341 pp.

McMaster, W. M. 1967. *Hydrologic Data for the Oak Ridge Area, Tennessee,* U.S. Geological Survey Water Supply Paper 1839-N.

Migdalski, E. C. 1962. *Freshwater Sport Fishes of North America.* (New York: Ronald Press) 431 pp.

Minckley, W. L. 1973. *Fishes of Arizona.* (Phoenix: Arizona Game and Fish Department) 293 pp.

Moore, W. G. 1966. "Central Gulf States and the Mississippi Embayment," in *Limnology in North America.* D. G. Frey, Ed., pp. 287-300. (Madison, WI: University of Wisconsin Press).

Moran, M. S. 1977. "Development of the Fort Payne Chert into an Aquifer along the Eastern Highland Rim, Central Tennessee," *Geol. Soc. Am. Abs.* 9(2).

National Academy of Sciences and National Academy of Engineering. 1972. *Water Quality Criteria, 1972.* (Washington, D.C.: U.S. Environmental Protection Agency) 594 pp.

Parker, B. C. 1976. *The Distributional History of the Biota of the Southern Appalachians. Part IV. Algae and Fungi: Biogeography, Systematics, and Ecology.* (Charlottesville, VA: University of Virginia).

Patriarche, M. H. and R. S. Campbell. 1958. "The Development of the Fish Population in a New Flood-Control Reservoir in Missouri, 1948-1954," *Trans. Am. Fish. Soc.* 87: 240-258.

Patterson, J. H.   1970.   *North America: A Geography of Canada and the United States*.   (Oxford, England: Oxford University Press).

Pennak, R. W.   1949.   "Annual Limnological Cycles in Some Colorado Reservoir Lakes," *Ecol. Monogr.* 19: 233–267.

Pennak. R. W.   1966.   "Rocky Mountain States," in *Limnology in North America*, D, G. Frey, Ed., pp. 349–370.   (Madison, WI: University of Wisconsin Press).

Peterson, D. D. and J. C. Sonnichsen.   1976.   *Assessment of Requirements for Dry Towers*.   HEDL-TME 76-82, Hanford Engineering Development Laboratory, Richland, WA.

Pierce, D. M. and J. E. Vogt.   1953.   "Method for Predicting Michigan-Huron Lake Fluctuations," *J. Am. Water Works Assoc.* 45: 502–520.

Price, D. and T. Arnow.   1974.   "Summary Appraisals of the Nation's Ground-water Resources — Upper Colorado Region," U.S. Geological Survey Professional Paper 813-C.

Reeder, H. O.   1978.   "Summary Appraisals of the Nation's Ground-water Resources — Souris-Red-Rainy Region," U.S. Geological Survey Professional Paper 813-K.

Reid, G. K.   1961.   *Ecology of Inland Waters and Estuaries*.   (New York: Van Nostrand Reinhold Co.) 375 pp.

Ricker, K. E.   1959.   "The Origin of the Glacial Relict Crustaceans in North America, as Related to Pleistocene Glaciation," *Can. J. Zool.* 37: 871–893.

Rima, D. R., M. S. Moran and E. J. Woods.   1977.   "Ground-water Supplies in the Murfreesboro Area, Tennnessee," U.S. Geological Survey Water Resource Investigation 77-86.

Shepherd, A. D.   1979.   *A Spatial Analysis Method of Assessing Water Supply and Demand Applied to Energy Development in the Ohio River Basin*.   ORNL/TM-6375, Oak Ridge National Laboratory, Oak Ridge, TN.

Sinnott, A. and E. M. Cushing.   1978.   "Summary Appraisals of the Nation's Ground-water Resources — Mid-Atlantic Region," U.S. Geological Survey Professional Paper 813-I.

Smith, S. H.   1972.   "Factors in Ecologic Succession in Oligotrophic Fish Communities of the Laurentian Great Lakes," *J. Fish. Res. Board Can.* 29: 717–730.

Stauffer, T. M. 1976. "Fecundity of Coho Salmon (*Oncorhynchus kisutch*) from the Great Lakes and a Comparison with Ocean Salmon," *J. Fish. Res. Board Can.* 33: 1144-1149.

Strahler, A. N. and A. H. Strahler. 1976. *Elements of Physical Geography*. (New York: John Wiley and Sons, Inc.) 469 pp.

Swindale, D. N. and J. T. Curtis. 1957. "Phytosociology of the Large Submerged Plants in Wisconsin Lakes," *Ecology* 38: 397-407.

Tanner, H. A. 1960. "Some Consequences of Adding Fertilizer to Five Michigan Trout Lakes," *Trans. Am. Fish. Soc.* 89: 198-205.

Taylor, O. J. 1978. "Summary Appraisals of the Nation's Ground-water Resources — Missouri Basin Region," U.S. Geological Survey Professional Paper 813-Q.

Terry, J. E., R. L. Hosman and C. T. Bryant. 1979. "Summary Appraisals of the Nation's Ground-water Resources — Lower Mississippi Region," U.S. Geological Survey Professional Paper 813-N.

Thomas, H. E. and D. A. Phoenix. 1976. "Summary Appraisals of the Nation's Ground-water Resources — California Region," U.S. Geological Survey Professional Paper 813-E.

U.S. Fish and Wildlife Service. *Federal Register*. Sept. 25, 1975; June 14, 1976; July 14, 1977; Aug. 11, 1977; Aug. 23, 1977; Sept. 9, 1977; Sept. 22, 1977; Nov. 21, 1977; Jan. 27, 1978; Mar. 27, 1978; Apr. 13, 1978; Apr. 18, 1978; May 23, 1978; Jan. 17, 1979.

U.S. Department of the Interior. 1971. *Proceedings of a Symposium on Rare and Endangered Mollusks (Naiads) of the United States*. S. E. Jorgensen and R. W. Sharp, Eds., Twin Cities, MN.

U.S. Department of the Interior. 1973. "Threatened Wildlife of the United States," Bureau of Sport Fisheries and Wildlife, Washington, D.C.

U.S. Department of the Interior. 1977. "Outdoor Recreation Action No. 43." Bureau of Outdoor Recreation, Washington, D.C.

U.S. Environmental Protection Agency. 1976. *The Influence of Land Use on Stream Nutrient Levels*. EPA-600/3-76-014, Office of Research and Development, Corvallis, OR.

U.S. Environmental Protection Agency. 1977*a*. "Coal Mining Point Source Category, Effluent Limitations Guidelines for Existing Sources," *Fed. Regist.* 42: 21388.

U.S. Environmental Protection Agency. 1977*b*. *National Water Quality Inventory/1976 Report to Congress.* EPA-440/9-76-024, Office of Water and Standards, Washington, D.C.

U.S. Environmental Protection Agency. 1977*c*. *Nonpoint Source-Stream Nutrient Level Relationships: A Nationwide Study.* EPA-600/3-77-105, Office of Research and Development, Corvallis, OR.

U.S. Environmental Protection Agency. 1977*d*. *North American Project — A Study of U.S. Water Bodies.* EPA-600/3-77-086, Office of Research and Development, Corvalis, OR.

U.S. Geological Survey. 1977. "Estimated Use of Water in the United States in 1975," Circular 765, Arlington, VA.

U.S. Geological Survey. 1970. *The National Atlas.* Washington, D.C.

U.S. Nuclear Regulatory Commission. 1977. "Draft Environmental Statement Related to the Determination of the Suitability of the Fort Calhoun Site for Eventual Construction of the Fort Calhoun Station, Unit No. 2," NUREG-0213.

Walton, W. C. 1970. *Groundwater Resource Evaluation.* (New York: McGraw-Hill Book Company).

Water Information Center. 1970. *The Water Encylcopedia.* Port Washington, NY.

Water Resources Council. 1974. "Project Independence Water Requirements, Availabilities, Constraints, and Recommended Federal Actions," Federal Energy Administration Project Independence Blueprint, Final Task Force Report, Washington, D.C.

Weist, W. G., Jr. 1978. "Summary Appraisals of the Nation's Ground-water Resources — Great Lakes Region," U.S. Geological Survey Professional Paper 813-J.

West, S. W. and W. L. Broadhurst. 1975. "Summary Appraisals of the Nation's Ground-water Resources — Rio Grande Region," U.S. Geological Survey Professional Paper 813-D.

Wild and Scenic Rivers Act. October 12, 1976, as amended through PL 94-486.

Wilson, J. N.   1958.   "The Limnology of Certain Prairie Lakes
in Minnesota," *Am. Midl. Nat.* 59: 418-437.

Woodruff, J. F. and J. D. Hewlett.   1970.   "Predicting and
Mapping the Average Hydrologic Response for the Eastern
United States," *Water Resour. Res.* 6(5): 1312-1326.

Wyoming Water Planning Program.   1970.   "Water and Related
Land Resources of the Green River Basin, Wyoming," Report
No. 3 (Cheyenne, WY: State Engineer's Office).

Yount, J. L.   1966.   "South Atlantic States," in *Limnology in
North America*, D. G. Frey, Ed., pp. 269-286.   (Madison, WI:
University of Wisconsin Press).

Zurawski, A.   1978.   "Summary Appraisals of the Nation's
Ground-water Resources — Tennessee Region," U.S. Geological
Survey Professional Paper 813-L.

Bull Shoals Reservoir 45
Burbot 66

Caddis fly larvae 75
Cadmium 19,36,42,46,61,69,77,
  82
Calcium 56
California State Water
  Project 100
Cambro-Ordovician aquifer 36,
  37
Canadarago Lake 12
Canadian River 45
Canadian Shield 23
Canals 19,57
Canyon Ferry Reservoir 68
Cape Fear River 19
Carbonates 67
Carolina Bays 18,21
Carp 27,34,56,66,75,76
Carpiodes 40
Carpsucker 76
  Also see Carpiodes
Carrizo Sand 43
Carrizo-Wilcox aquifer 52,53
Carson Sink 84
Cascade Range 88
Catawba River 19
Catfish 27,34,75
Catskill Mountains 11
Cattail 6
Center Hill Reservoir 28
Central Arizona Project 84
Central Basin 14
Central Coastal subregion 88,
  100
Central Lowland 46
Central Utah Project 86
Central Valley 88,100
Central Valley Project 100
Centrarchids 12,21,40,50,56,76
Charlotte, North Carolina 20
Chattahoochee River 18
Chattanooga, Tennessee 13
Chehalis River 101
Chesapeake Bay 7,11
Chicago, Illinois 26
Chickamauga Lake 12
Chironomids 40,66,67
Chloride 29,36,56,67
Chromium 36

Chubs 27
Cimmaron River 45
Ciscoes 66
Cladocerans 27
  Also see Daphnia
Claiborne Group 43
Clark Fork River 101
Clark Hill Reservoir 18
Clear Lake 87
Clinch River 12,13
Coachella Canal 87,100
Coachella Valley 87
Coal 33,63,75
  mining 9,12,13,29,34,41,76
Coastal Plain (including the
  Coastal Plain province) 6,9,
  11-13,18,19,21,46,52,53
Coast Ranges 88
Cockfield Formation 43
Coconino Sandstone 83
Coeur d'Alene Lake 101
Coffee Sand 13,42
Colorado desert 88
Colorado Desert subregion 88,100
Colorado Plateau 56
Colorado River (California WRR)
  100
Colorado River (Texas-Gulf WRR)
  51,54
Colorado Rockies 77
Columbia River 26,101,103
Conemaugh River 34
Connecticut River 3,4
Connecticut River Valley 4
Copepods 27,66,75,81
Coregonids 27
Crappie 17,66,76
Crater Lake 101
Critical habitats (defined) 105
Cumberland Plateau province 14,
  15,29
Cumberland River 28
Cuyahoga River 23
Cuyama River 87
Cyprinids 21

Dakota Sandstone 72,78
  aquifer 61,78
Dale Hollow Reservoir 28
Dallas-Fort Worth, Texas 52

122